D0778757

little

free

library

NORTHERN LIGHTS LIBRARY SYSTEM

Scan me

The Good Cook's Book
of Salt & Pepper

Other Books by Michele Anna Jordan

More Than Meatballs

The Good Cook's Journal

The Good Cook's Book of Mustard

The Good Cook's Book of Oil & Vinegar

The Good Cook's Book of Tomatoes

Vinaigrettes & Other Dressings

The World Is a Kitchen

Lotsa Pasta

VegOut! A Guide Book to Vegetarian Friendly Restaurants in Northern California

The BLT Cookbook

San Francisco Seafood

The New Cook's Tour of Sonoma

Pasta Classics

California Home Cooking

Polenta

Pasta with Sauces

Ravioli & Lasagne

A Cook's Tour of Sonoma

The
Good Cook's Book
of Salt & Pepper

ACHIEVING SEASONED DELIGHT, WITH MORE THAN 150 RECIPES

MICHELE ANNA JORDAN

Photography by Liza Gershman

Skyhorse Publishing

Copyright © 1999 by Michele Anna Jordan
New material copyright © 2015 by Michele Anna Jordan
Photographs copyright © 2015 by Liza Gershman, except where noted

Originally published as *Salt & Pepper* in 1995 by Broadway Books.

All rights reserved. No part of this book may be reproduced in any manner without the express written consent of the publisher, except in the case of brief excerpts in critical reviews or articles. All inquiries should be addressed to Skyhorse Publishing, 307 West 36th Street, 11th Floor, New York, NY 10018.

Skyhorse Publishing books may be purchased in bulk at special discounts for sales promotion, corporate gifts, fund-raising, or educational purposes. Special editions can also be created to specifications. For details, contact the Special Sales Department, Skyhorse Publishing, 307 West 36th Street, 11th Floor, New York, NY 10018 or info@skyhorsepublishing.com.

Skyhorse® and Skyhorse Publishing® are registered trademarks of Skyhorse Publishing, Inc.®, a Delaware corporation.

Visit our website at www.skyhorsepublishing.com.

10 9 8 7 6 5 4 3 2 1

Library of Congress Cataloging-in-Publication Data is available on file.

Cover design by Erin Seward-Hiatt
Cover photo credit Liza Gershman

Print ISBN: 978-1-62914-576-1
Ebook ISBN: 978-1-62914-966-0

Printed in China

for
Gina and Nicolle: *I love you both more than salt, and pepper, too.*

Table of Contents

Part III: A Salt & Pepper Cookbook

Part IV: Appendices

Acknowledgments

First, I must acknowledge writer Sallie Tisdale, whom I do not know. Her remarkable book *Lot's Wife: Salt and the Human Condition* is the finest treatise I have read on salt. Her writing, so rich with passion and curiosity, is a continuing inspiration, as is Marilynne Robinson's exquisite novel *Housekeeping*, which I have quoted in this book.

Many thanks to Harriet Bell, my former editor, and Doe Coover, my former agent, for believing in this book long before salt had caught fire. And to my current editor, Andy Ross, mahalo for finding it a new home. Nicole Frail and the team at Skyhorse Publishing get a huge shout-out, too, for shepherding my beloved little book through production. *Thank you so very much*; I am both grateful and humbled.

There are no words in English to adequately thank photographer Liza Gershman and the team of volunteer assistants who worked on a crazy schedule to help produce Liza's lovely images. Fabiano Ramaci, Rayne Wolfe, Kelly Keagy, Sherry Soleski, Clio Tarazi, and Deborah Pulido, you all offered invaluable support and I offer an enthusiastic grazie!

Cultivate, a sweet little cookware store in downtown Sebastopol, loaned many props for our photo sessions. Thank you! And thank you for your kindness and support of my books. Thanks, too, to K & L Bistro, a fabulous family restaurant in downtown Sebastopol, for the marrow bones (twice!) and for always taking such good care of me when I stumble in after too much work.

Marty Goldsmith, vice president and senior trader at Ludwig Mueller Co., Inc., in New York has generously shared his vast knowledge of the world of pepper for both editions of this book. Thank you, Marty!

Tourism Malaysia provided crucial support and introduced me to a region of the world I might not have discovered on my own. A heartfelt *terima kasih* to Lily Musni and Sharifah Danial in the Los Angeles office, Raja H. J. Normala and her associates in the Kuala Lumpur office, and Talib H. J. Long in Kuching, Sarawak. Many thanks to Malaysia Airlines, too.

Had it not been for Bill Penzey of Penzeys, Ltd., I might not have gone to Malaysia at all, so Bill, *thank you.* (But would you please get Naturally Clean Black Pepper back in stock!)

In Malaysia, so many people were kind and helpful that I am sure I will forget to name them all, and I offer my apologies in advance of my thanks. That said, thank you very much to Elizabeth Foo of the Legend Hotel in Kuala Lumpur, to Fiona Fong of the Holiday Inn in Kuching, and to Anandan Adnan Abdullah, former general manager of the Pepper Marketing Board and a gracious and charming host. It was through Anandan's efforts, and those of his assistants Chong Vui Hok and Joseph Lau, that I was able to return to Sarawak for a second visit not long after my first. I have fallen completely in love with the city and the state of Sarawak, and the island of Borneo. And to Narain Nallathamby, who was my guide in Kuala Lumpur, thank you for indulging my whims and curiosity, and for being a wonderful companion. And to the sweet Ramlee Eli, hugs and fond memories.

Thanks to everyone who helped along the salt trail, especially Jill Singleton and Skip Niman of Cargill Corporation; Andy Briscoe of the Salt Institute; Brian Abendroth, and the folks at the Grain and Salt Society. And to Elizabeth Erman of ASTA, thanks for leads to pepper traders. Special thanks go out to the Poland Tourism Board, too, for whisking me off to the Wieliczka Salt Mine so soon after I'd finished the first edition of this book.

And heartfelt thanks and love to my daughters Gina and Nicolle; my grandson Lucas; my son-in-law Tom; my dear friends John Boland and James Carroll; my friend Ken "Poco Torta" Behrens; my childhood friends

Bobby and Connie Howard and Linda Zalesky; my kumu Shawna Alapa'i and all my hula sisters and my dear friend Mary Duryee. Finally, to the DeLaura family, I offer my deepest *mahalo nui loa* for having you back in my life.

And to Patrick Bouquet, wherever you may be, I miss you more than salt, every minute of every day.

Introduction
to the Second Edition

When I wrote the first edition of this book, I thought it was a year or two ahead of the curve. I believed the salt wave was about to crest and would soon crash across our tables in a rush of deliciousness.

I was wrong. It was to be a decade before the frenzy for signature salts reached its crescendo. By the time it happened, my little treasure chest of a book was all but forgotten. Author Mark Bitterman rode the wave like a champion surfer bathed in a salty mist. His book *Salted* (Ten Speed Press, 2010) shows 157 different salts and he has carved out a unique position for himself in this savory world. He is a salt sommelier or *selmelier* at The Meadow, which he co-founded. The shop—there are three locations and an online store—sells more than a hundred premium salts. *Good job, Mark!*

A dozen or so other books on salt, including a second one by Mark, have appeared since *Salt & Pepper* vanished from retail shelves.

Specialty salts are now everywhere or nearly so, readily available to chefs, home cooks, and anyone eager to explore the planet's salty heritage. There's even a salt named after Sonoma County, even though it has no physical connection to the glorious place where I am lucky to live. It is more like the GMC Sonoma truck, inspired by a name that much of the country views romantically, as an Eden, a paradise of good living and all that it involves. This perception is, to a large degree, accurate but that is a story for another time. One of the things we do not produce here is commercial salt. We don't grow peppercorns, either.

As salt—artisan salt, finishing salt, condiment salt, salt tasting bars—has become all but ubiquitous, pepper has not enjoyed similar attention. There is no peppercorn frenzy, no other book that explores its history. There are no peppercorn sommeliers (though as this book goes to press, New York City has its first hot sauce sommelier).

Indeed, it has become harder—all but impossible, really—to get some of the world's finest peppercorns. Almost no Malaysian pepper is imported by U.S. supplies and the best, Naturally Clean Black Pepper and Creamy White Pepper from the Malaysian state of Sarawak on the island of Borneo, is unavailable in the United States. Most retailers don't know about it and many don't understand what it is or why it is so good. It is almost as if this book's chapter on pepper, in which I speculate about what would happen if pepper vanished, was prescient. The finest pepper in the world is, in this country, little more than a rumor, a tale of deliciousness that most people don't believe. *So what?* They shrug.

My love of this pepper is not an act of faith. It is a fact. I have about half a jar left of the fifty plus pounds I've purchased and when I think I'm exaggerating, when I fear my romantic nature has once again gotten the better of me, I open the jar and breathe in the peppercorn's beautiful floral aromas and sigh. It is so pretty. Maybe the Pepper Muse will find me and whisk me off to Kuching. I'll travel lightly, fill my bags with pepper for my return and I promise to share with you, dear reader.

Introduction
to the First Edition

Salt seasons the self so that the self's own true flavor emerges.
—Sallie Tisdale, *Lot's Wife: Salt and the Human Condition*

You know me. I'm the shy girl from elementary school who gave you her ice cream, the one who scraped the chocolate frosting off the birthday cake and *ate the cake*. Take me out for a drink and you have to keep an eye on me or I'll lick the salt off your margarita glass.

My friend John loves desserts, looking forward with relish to the end of the meal and its last, sweet pleasure. I look across the table and appreciate John's delight, though such pleasure is mostly unavailable to me. I have little interest in sweets. I was born without a sweet tooth and I am an outsider, my nose pressed up against the window of another's joy. Certainly, I can appreciate a voluptuous cheesecake, admire a luscious chocolate mousse, and sincerely believe that tarte tatin is one of the great contributions France has made to world culture. But I am almost always satisfied by a single bite. I don't have a chocolate jones. Ahh, but that salty flourish. I crave it.

When I was a little kid, I'd sneak into the kitchen and find lemons to squeeze on my hands. Then I'd sprinkle salt over the juice, sit in front of the television, and slowly lick off the tangy film. If I was in the mood for a more elaborate snack, I might peel a lemon, cut it into the thinnest slices, and sprinkle the slices with salt, a pleasure I gave up only in my early twenties, when a friend warned that the acid would eat the enamel off my teeth. He knew; it had already happened to him.

This book began more than a decade ago as self-defense. The salt police were everywhere, ripping shakers out of hands and wagging fingers at the loosely indulgent, me among them. I took them to task in my column "The Jaded Palate," published then in the *Sonoma County Independent*. The response was encouraging, and I was buoyed by the realization that I was not alone in my salty sea.

The column soon grew into a book proposal, something I worked on late at night when I was done with my serious assignments. I kept it secret for years. Then articles began appearing in other publications, first in the *New York Times*, then in the *Washington Post*, and suddenly everywhere. My secret passion had become grist for the mill; I worried that I had missed my chance. I sent the proposal to my agent Doe Coover, but she said nothing, nor did I. You never knew where the salt police might be lurking. Maybe she thought I had slipped off the shore: a book on salt and pepper?

A few months passed and I saw another article. "Sell it," I said, and she did, almost instantly. That was two, maybe even three, years ago and the fervor for salt grows daily. Nor should we ignore its sultry spouse, pepper, the sneaky one of the duo; interest in it, too, has been on the rise. But salt is always stealing the spotlight from pepper. Salt tap-dances naked on your table, makes you blush with delight. Pepper taps you on the shoulder and invites you behind closed doors. Both are shameless in the endless pleasure they impart.

As I was leaving New York for San Francisco one winter morning, my dear friend Peter handed me a small present.

"It's for your Muse," he said, aware that I was heading home to begin the book in earnest. It was an exquisite gift, one of the most precious I have ever received, a crystal cross carved of salt, carried all the way from the Wieliczka Salt Mine in Poland. I wrapped it carefully in a small cloth and carried it on the plane, tucked into the pocket of my velvet jacket, where I ran my fingers over its hard smooth surfaces every now and then. I never licked it, I swear.

I have since become a magnet for salt. I have a tiny wooden salt box full of a fragile flakes, salt that a friend gathered from a natural brine pool off the Sonoma County coastline. In my kitchen, there's a tiny pile of salt made by a photographer who collected saltwater near her home in Maine, boiled it to concentrate the brine, then waited patiently for the liquid to evaporate and the crystals to form. I awake to find gifts of salt on my porch; stories about salt arrive by fax and email and post (one in Japanese, which I cannot read). My daughter Nicolle brings me a present, *Salt Hands*, the story of a deer who came to lick salt out of a young girl's hands.

I worked all summer (but who could tell the time of year, here in my quiet study where I am surrounded by salt and pepper, away from the sun) to complete the manuscript. One day a phone call came: *Would you like to join a press trip to Poland?* an unfamiliar voice inquired. Yes, Wieliczka is on the itinerary, I am told, I shiver. It is the Salt Muse, calling me home. As I print out the last pages of *Salt & Pepper*, I have just enough time to pack.

PART I
Salt

Magical Salt

"I love you more than salt," the mythical young princess says to her father. Her older sisters, articulate in their hyperbole—more than gold, they tell the king, more than my own life, as much as God—are rewarded with riches and kingdoms of their own. The arrogant king, devastated to be so poorly valued by his favorite child, condemns her to a life of loneliness and poverty in exile.

The king's next meal is flat and tasteless, and so is the next, and the next, and the next. For thirty days, his food lacks flavor. You might conclude that it is sadness and loss that eclipse his pleasure, but it is not. He summons his chef and demands an explanation.

"You value salt so little, your majesty," the chef explains coyly, "that I no longer use it in your food."

You can see where this is going. The princess is returned from exile, reunited with her secret fiancé—the wily chef of course—and given a lavish wedding and riches that far surpass those granted her sisters. The king's food sparkles with flavor and savor once again.

I was told the story by a cooking student, a woman from India who remembered it from childhood, when she had likely heard it from an English nanny; there are many versions of this tale throughout Europe. Common salt is more essential than we like to admit, each version cautions us.

Another scene: It is Halloween, the eve of el Dia de los Muertos, and a young Mexican girl is boiling an egg. After it is cooked, she runs it under a cool spigot, holding the hot egg even as it burns her impatient hands.

When at last it is cool, she removes the shell and carefully breaks the egg in half, revealing the round yellow yolk, the kernel of life at the core, which she quickly discards. She lifts the lid of a nearby box of salt and takes as big a pinch as her fingers can hold, depositing it in the hollow of the egg. She continues, pinch after salty pinch, until the center of the egg is filled. She fits the halves together and tiptoes silently to the bedroom she shares with her sister, where she awaits the stroke of midnight.

Eventually the hour comes, signaled by distant chimes, and she puts the egg with its hidden seed, its talismanic treasure, into her mouth, chewing and swallowing quickly so that the jewel of salt does not burst and leave her parched, for she must not take a drink. Her sister wakes and calls her name, but she pretends to be asleep. To speak would break the spell. Who will it be? Who will appear in her dreams and offer her a quenching drink of water? Her life's mate, the story goes, and she falls asleep imagining his face.

Ah, magical salt. It seasons our tales, and spices up our language, each word, every phrase a nod to a single truism: Life is tasteless without salt.

"My little salt box," a pretty young girl in Andalusia might be called by her suitor.

"The salt of the party," Arabs say.

Early Christians rubbed newborn babies with salt. Asians, confronted with a dull youth, shake their heads and whisper, "He was not salted when he was born," a statement so precise and so evocative that I believe we should adopt it immediately.

"Somebody forgot to salt the popcorn," I imagine one friend whispering slyly to another as a third bores them to tears.

To say there is salt between us indicates a bond that must be honored. Trespass not against the salt, the Greeks warned. In Iran, to be disloyal or ungrateful is to be untrue to salt. While we were visiting the salt mines in Wieliczka, Poland, Mariusz Moryl, marketing manager for the Polish National Tourist Office, told me that in his country old friends say they've eaten a lot of salt together. Those same friends might complain about a salty bill when the tab at a restaurant or pub is higher than expected. An indolent employee has long been described as not worth his salt, a reference to Roman soldiers who were paid a portion of their wages—that is, their salary—in salt or with extra currency to buy salt. Salt of the earth surely is a compliment, if a rather patronizing one these days.

The spilling of salt is a bad omen, both literally and figuratively. "The Romans are said to have led victims to execution with salt balanced on their heads," Sallie Tisdale tells us in *Lot's Wife*. "When it spilled, so did their blood." Spilled salt foreshadows loss of friendship, a broken heart or bone, a shipwreck, a death. Throw a little over your shoulder, my mother warned me whenever I knocked over the salt shaker. This ritual, so automatic to so many of us, began as an appeasement to the demons said to hover at our left, awaiting a moment of weakness that would give them access to our souls, clumsiness apparently a chief means of entry. Throwing salt, it seems, stopped them in their invisible tracks.

It was European Christians who figured this out; salt, for a time, was their domain. Alleged witches, thought to be worshipers of Satan, were said to hate it; a meal without salt, an unholy affair, became known as a "witch's supper." Heavily salted foods were thought to ward off demons.

In recent times, we have believed exactly the opposite. Salt, we have been cautioned, is an open invitation to the demons we fear the most: high blood pressure, death by stroke and heart attack, cancer. For decades, salt has been our demon lover, our secret culinary paramour, the kinky pleasure we pretend not to crave. But it's a ruse. We have always craved salt.

The story of salt and our craving for it is inseparable from human history. After we crawled out of the sea and wandered inland, we were able to do without extra salt for a time because we got what we needed from the raw meat we ate. But as we grew more and more civilized, cooking our food and learning to farm, our need for salt began to rise, growing into an insatiable hunger if unsatisfied for long.

This hunger must have led us to the nearest salt licks, likely already polished smooth by the tongues of a thousand zebras, giraffes, elephants, cheetahs, deer, and elk. Perhaps we followed them to the salt licks, as we later followed the buffalo of North America to natural outcroppings of salt. We've learned something in the process: Today, we leave manmade salt licks in the wild and build observation tents on hills above them, knowing that eventually the wild beasts will come for the salt. (Don't worry, we want to see the animals, not hunt them. It is illegal to place salt licks in areas that allow, say, deer hunting.)

Those of us who obeyed salt's call lived to reproduce and passed our salt literacy on to our children. Those of us without a taste for salt, those of us who didn't care, died out eons ago. You wouldn't be here if your ancestors had forsworn salt. What is this salient thing, salt, that we need so badly?

Salt is a rock, an inorganic mineral composed of 40 percent sodium and 60 percent chloride, joined by one of the strongest chemical unions there is, an ionic bond. It is a perfect fit; the molecule is often used in chemistry texts to illustrate the principles of attraction and union. When a salt crystal grows without interference—that is, without interlopers (magnesium, calcium, all those other substances that might go along for the ride) intruding on the sodium and the chloride—it is perfectly and exactly square.

Sodium, dissolved in the extracellular fluid that assists with the transportation of nutrients, is one of the main components of the body's internal environment. Sodium functions as an electrolyte, as do potassium, calcium, and magnesium, all of which regulate the electrical charges within our cells. Chloride supports potassium absorption, and helps oversee the body's acid and base balance. It enhances carbon dioxide transportation, and is an essential component of digestive acids. We need it, and so we crave it. It's that simple, and that complex.

What Is Salt, Anyway?

Salt is a simple chemical made up of just two components. Salt, that is to say, sodium chloride or common salt, the chemical symbol for which is NaCl, forms when sodium, a soft metal, and chlorine, a gas, share electrons in what is called an ionic bond, one of the strongest chemical bonds there is. Once these two elements share electrons, they become magnets, the sodium positively charged and the chlorine negatively charged. The magnetic force holds the sodium and chlorine together in an intricate array of sodium ions surrounded by chlorine ions in turn surrounded by sodium ions, ad infinitum. It takes a great deal of force to separate the two once they have joined. Dissolving salt in water does not loosen the bond, nor does heat or time erode it. The specific configuration of these ions, which creates four 90-degree angles in each molecular bond, and their endless repetition create the perfect crystalline structure of salt, and account for the fact that salt can be cleaved into ever smaller crystals that retain their smooth, angular surfaces. The enormous crystals that can be seen in salt museums are called halite, a term that accurately applies to all rock salt. Some halite seems to glow with colorful inner lights, usually blue from suspended cobalt and red from suspended iron. It was rock salt colored in this way that is said to be the reason that the ground near Sodom turned red just before Lot's wife looked back and met her fate.

Salt's Mystery

"Salt," I said, writing to Harold McGee, "heightens the flavors of virtually everything, but I'm hoping to answer the question of exactly why and how it does this. Is it enough that salt draws out the moisture and thus the flavor of an ingredient, and then, as we eat, additional salt dissolves slowly on the tongue, thus bringing flavors to the palate and creating a harmonious finish among them?"

"I'm afraid that your question is indeed more complicated than it seems," the well-known food scientist and curious cook responded in an email message written in France. The function of salt "in fact is the subject of much ongoing research. Pretzels and chips and freshly salted foods aside, the salt in food is already dissolved . . . and in equilibrium with the water and other constituents. Since there are only a few genuine 'tastes'—sensations registered on the tongue, not the nose—certainly salt contributes to the overall complexity and balance of flavor.

"In addition to that, the concentration of salt in a food helps determine chemically how many of the other flavor components are going to behave: that is, how available they are to our senses. But this effect varies among the many hundreds of flavor components in a given food, so it's hard to generalize: except to say that salt contributes more than saltiness to flavor.

"Sorry there's not a neater answer, though maybe it's good to retain some mystery in such a basic matter."

Taste, like love, defies reduction. And just as love is indispensable to the savor of life, so salt is essential to the enjoyment of food. Salt is flavor's midwife; grains scatter across our tongues and melt like tiny stars, enchanting and mysterious, inseparable from taste itself. Savory foods do not reach their full flower without its skillful application; sweets blossom with a judicious sprinkle of salt crystals. Professional chefs identify salt as the single most important ingredient in their kitchen and insist that their new sous chefs learn immediately how to salt properly the foods they cook.

Common Salt, Common Sense

Today, we believe that we eat much more salt than we ever have. "It's all that processed food," people say, usually preening with superiority as they add that they never use salt. Ho hum. I feel sorry for them, and wonder what else besides good food they deny themselves, but I try to keep quiet, unless too many of them appear at once. Then I get awfully cranky and start citing statistics and talking about Americans' annoying bias against pleasure. I lost my patience with the salt police about the time a boyfriend took the shaker out of my hand and told me that, henceforth, *he* would decide what foods I would be allowed to salt. (The next one considered putting a salt lick next to the bed, to keep my spirits up, he said; I kept him around a bit longer.)

Yes, we certainly have a penchant for munching salty snacks, but in fact we consume much less sodium chloride than our ancestors did. In the Middle Ages, foods were bathed in salt. Meat and fish preserved in it were the common centerpieces of a meal—this was before refrigeration, before manmade ice, how to preserve the kill, if not with salt?—and more salt was added at the table, cascading off the tips of knives dipped into the elaborate salt cellars that graced the tables of the wealthy. (Status, it should be noted, was revealed by where one was seated in relationship to the salt cellar; to be "below the salt" was an intentional insult or an indication of low status.) Twenty grams a day—about two tablespoons of coarse sea salt—or more were commonly eaten.

In the past century and a half, our consumption of salt has remained fairly level—between about six and eleven grams (between 1¼ and 2¼ teaspoons of table salt) a day is the world average—with little if any fluctuation in response to the good news and bad news that has been circulating, like a salty brine, for decades.

Recommendations by the FDA have recently been adjusted downward, calling a teaspoon or even less the most one should consume daily.

A cookbook won't put the controversy to rest, but I must add my voice to the choir. I simply can't conclude that anything as essential to our well-being as salt, anything that provides such primal pleasure and delightful satisfaction, is bad. To believe this would be to believe the world itself is made wrong. I believe the world is right and wise (not to mention delicious), and that our true cravings lead us where we should go. There are a thousand ways we can eclipse our body's wisdom, but, still, it cries out for salt.

"The amount of sodium ions [is] so critical," according to Thomas Moore, writing in *The Washingtonian*, "that the concentration of dissolved salt in the blood is not allowed to vary by more than 1 percent." If everything is working as it should, our kidneys serve as the regulators of salt: Eat more than we require, and more is excreted; eat less, and our bodies, via our kidneys, hang on to what we already have.

Our bodies hear salt's call and respond with longing if something goes haywire. Salt craving—a desire that drives salt-starved children to dip every morsel of food into a pile of salt, and can compel us, in *extremis*, to eat dirt to extract its smidgen of salt—is the body's way of acquiring life-sustaining salt. If we cannot hold on to our salt, if our body squanders it in excretion, an overpowering craving will eclipse every other impulse; we will become our body's salt servant, or we will die.

Studies completed in the late 1980s and early 1990s indicate that, indeed, salt is not the killer it has been declared to be. The war against hypertension, which began in the 1930s and kicked into high gear during the Nixon administration in the early 1970s, attacked salt; for decades it has been considered a major culprit, declared guilty without a fair trial. Yet new studies show an increased risk of death among those with the lowest sodium levels, and they show a large population (between 75 and 80 percent) unaffected by salt. (Salt yea-sayers point to Japan, where consumption is about double ours. Life expectancy there is eighty years; here it is seventy-six.) And no studies show that salt increases blood pressure; rather, certain studies demonstrate that some hypertensives (about 8 percent of the general population) can reduce an already elevated blood pressure by reducing the amount of salt they eat.

The vast majority of people who drink alcohol do not become alcoholics, or even problem drinkers. Some do, but does this mean that everyone should become a teetotaler? Likewise, if some people with hypertension can reduce their blood pressure by reducing salt, should we all follow suit? There are those who would answer yes to both questions—better safe than sorry, they argue, and with straight faces—but scratch a zealot and you'll find a Puritan singing to the tune of, "if it tastes good, it must be bad." Of course advice given to a symptom-specific group should not apply to the general population. Common sense and common salt should join hands, make up and go steady, reignite their romance. *Come over here, give me a salty kiss.*

You may already suspect my conclusion: If you have a medical condition that requires you to limit the quantity of salt you eat, that is indeed unfortunate, but it is not a reason to impose similar restrictions on everyone else.

There is yet another side to this controversy. Salt, once so dear, is in modern times plentiful and cheap, easily taken for granted. Having salt was once proof of status; today to abstain is proof of virtue, of taking the high road. Since the early 1980s, many Americans have boasted that they "never eat it," as if those of us who do are somehow lax, indulgent, common, as common as salt. "I love you more than salt" falls on deaf ears. *So what?* they think.

Yet these same salt snobs seem to forget about the fast foods on which they rely, the pickles, popcorn, and lunchmeats, the Big Macs and Whoppers, the prepared entrées from the freezer compartment of the neighborhood market. Most prepared foods are laden with salt because food manufacturers know an essential detail of its nature: Salt dissolves only in the desire for more. It makes otherwise bland foods taste good, and it makes good foods taste even better. Its absence can render potentially wonderful things tasteless. Given the hysteria of recent times, we might have seen a proliferation of low-salt and no-salt packaged foods, but you know what? They don't sell, because no one likes them. They come and go on the market shelves but none last for long.

If you need to cut down on salt, where do you think you should start? By denying yourself a sprinkling of tasty crystals on your summer tomatoes? By refusing to salt the pasta water, or the grilled eggplant, or that lovely batch of pesto you have just made? Or by leaving lean cuisine entrées and bottled salsa on the grocer's shelf? You know my answer.

Iodized Salt

As you have no doubt noticed, much of the salt sold in supermarkets is labeled "iodized," which indicates that the salt has had iodine added to it, usually in the form of potassium iodide. This practice began in the 1920s and was developed, in response to research by David Marine and his staff at the Michigan State Medical Society, to prevent thyroid goiter, which was epidemic in certain regions of the United States and particularly among children in the Midwest. The program to encourage families to use iodized salt spread throughout the country, and within thirty years, 76 percent of American households used only iodized salt. The epidemic of goiters virtually disappeared. Today, more than half of all table salt sold contains the micronutrient. Seafood as well as unrefined sea salt contains iodine naturally, and the supplement is not necessary if there are sufficient quantities of either in one's diet; we require less than 225 micrograms of iodine a day. But in those developing countries where seafood is rarely if ever eaten, iodine deficiency remains a serious health problem today. In 1990, the World Summit for Children named it their top health priority and stressed that iodized salt is still the most efficient means of combating the deficiency.

Governments all over the world have attempted to impose a heavy tax on salt. The French salt tax, the *gabelle*, wasn't repealed until after World War II, and at least one writer blamed the dearness of salt in France for the country's falling birthrate (cows had recently been shown to suffer compromised fertility when there was insufficient salt in their diets). Today, salt is treated as a food product and as such is rarely taxed.

The Flowering Sea

I stand on the edge of a vermilion pool, a saturated virgin pickle whose frothy shores resemble pink rock candy. Five years ago, this was saltwater

Photo courtesy of Cargill Incorporated.

in the San Francisco Bay; today it is a soupy scarlet brine—colored by the algae that bloom red at the specific salinity of this pond—the last stage before it becomes salt. After a summer of wind and sun, most of the salt will have crystallized.

On the midsummer day of my visit, the calm crimson pool is speckled here and there with tiny seed crystals, California's equivalent of fleur de sel, though it will not be harvested as such. Instead, larger crystals will grow on the seeds of salt until they fall of their own weight to the bottom of the shallow pool, a foot, sometimes two feet, deep. In the fall the salt will be harvested—in a good year, nearly a million tons of it from seventeen hundred acres of red crystallization ponds—and added to a nearby salt mountain, ninety feet high and nearly a thousand feet long, to await further processing. Eventually, it will be washed in a saturated brine, redissolved, dried in a kiln, and sent through a mill to crack it to size. By next year, it will be sold as table salt, sea salt, iodized salt licks, rock salt for softening water and melting icy roads, and industrial salt, all of it from this same sea source.

Salt   **11**

Salt has a long history in San Francisco Bay. For centuries native Ohlone tribes gathered it from natural brine pools along the southern end of the bay until the Spanish explorers came along, built their missions, and declared the salt (and almost everything else) their own. In 1854, a German immigrant known as John Johnson founded California's modern salt industry; it is one of the state's oldest continuously operating industries.

Salt processing was lucrative when there was a single producer; Johnson's salt was soon fetching fifty dollars a ton, but it was, of course, only a matter of time before competitors came along. Less then twenty years after Johnson founded the industry, there were eighteen salt companies along the bay, a number that declined as prices dropped in response to overproduction.

From 1936 until its sale to Cargill Incorporated in 1978, Leslie Salt Company was the dominant producer, with holdings of forty thousand acres of salt ponds ringing the bay. Cargill maintains the familiar name

Barefoot Farmers

French sea salt may be sexy, but there is another side to the story of the romantic salt farmer. An Indian folk song moans, "Oh, mother, why did you marry me to a salt worker?" A salt farmer in India has a hard life. Salt is cheap and a family may spend several days harvesting a single ton of salt, which in 1998 brought less than four dollars. In the monsoon floods of that year, thousands of salt workers and their families, living in fragile shacks along the coast near Kandla in the state of Gujarat, were killed by a tidal wave, a cyclone, and subsequent floods. Even in the best conditions it's a dangerous occupation; the harsh Indian sun reflecting off the white mountains of salt is hard on the eyes. Constant exposure to the salty brine can lead to skin lesions which can become gangrenous—when there's too little salt, wounds can't heal; when there's too much, they won't. What are those Frenchmen thinking, not wearing shoes?

—*Time* magazine, 29 June 1998

of Leslie as a brand, along with nearly a hundred other custom-labeled brands, from Hain Sea Salt to Safeway Iodized Table Salt, all the same salt from the same source, but with various additions made according to the customer's specifications. The company has reduced the area devoted to salt production (but not the yield, because of improved crystallization and harvesting techniques), and sold and donated ten thousand acres for marsh restoration to the California Department of Fish and Game.

Planes at San Francisco International Airport roar above the glassy patchwork of saline ponds, and when passengers glance down on the green, blue, and blue green, rust red and bright red, slate gray and pitch black ponds, their gaze rarely lingers. Industrial salt, they might think, if they think of salt at all. Who cares about California salt, no matter how beautiful the ponds that are its source? It's French salt that's sexy.

For cachet and for mystique, it's hard to beat *fleur de sel*, flower of the sea. As a commercial product, fleur de sel is a modern conceit, a product of northwestern France—Île de Ré, Noirmoutier, and Guérande are the regions of production—virtually unheard of before the 1980s. It is a by-product of sel gris, the gray sea salt that in the late 1990s became so popular.

Fleur de sel blossoms on the surface of the saline pools that give us gray sea salt. When the conditions are just right, when the sun is hot and the wind is up, the salt blooms, crystallizing out of a saturated brine. Nearly instantly it forms a fragile skin over the salty liquor. For decades, the sauniers (salt farmers) broke up the veneer of crystals as they raked the bed of salt; left intact, it forms a shield that slows evaporation. But by the 1980s, French chefs had discovered fleur de sel and farmers were only too pleased to fill the new demand.

In the late 1990s, fleur de sel was the world's most expensive salt, commanding prices in the United States of more than thirty dollars a pound. Since then, as the fever for specialty salts has soared, that is no longer true, though some small purveyors sell fleur de sel for nearly double what it was fetching in 1998.

Salt farming has become a popular profession in France. In each of the regions, there are new training schools with more applicants than positions. In America in the early 1970s, young people headed to rural areas to live closer to the land and escape increasingly corporate sources of food. In France, young people flock to the West Coast for the same reasons, for the innocence and rhythm of farming salt. It adds to the perceived romance of the salt itself; photographs show beautiful young men (Frenchmen, at that), clad in nothing more than denim shorts, long hair tousled by the coastal winds, raking piles of glittering crystals from aquamarine pools. *What about that salty kiss?*

Condiment salt. It was inevitable that the term would be coined, and coined it was. I don't know who first used the phrase, but by the mid-1990s salt lovers, salt sellers, and salt farmers were all using it to refer to *fleur de sel* and a few other specialty salts. When Mark Bitterman of The Meadow, a boutique of artisan products founded by his wife, published his highly successful book *Salted* in 2010, 52 of the 157 salts he describes are recommended solely to finish a dish, which is to say as a condiment. As the country's official stance on salt continues to be "less is better," a lot of us have gone salt crazy.

Fleur de sel remains the premier condiment salt to all but a few passionate aficionados, like aceto balsamico tradizionale or the finest extra virgin olive oil. The cream atop the milk, it has been called, the caviar of salt. Its flavor is delicate, yet full and round in your mouth; it does not sear the tip of the tongue as some salts do; there is no bitterness, no sharpness. The most important characteristic is its texture; it crunches pleasantly between your teeth, and because it is crystalline rather than flaky, it dissolves slowly: Sprinkled on summer tomatoes, it lingers, salty jewels for the tongue to savor.

If fleur de sel leads the pack of artisan salts when it comes to reputation, *sel gris* isn't far behind. As late as 1996, it was a well-kept secret, praised by food writers, coveted by chefs, and loved by anyone who had visited the

Salt Tasting in Wine Country

When Ernie Shelton of Shelton's Market in downtown Healdsburg, California, saw a huge array of specialty salts at a natural foods expo several springs ago, he did not hesitate.

"Nobody is doing this," he said to his two brothers, "let's build a salt bar, the first in wine country."

Ernie has worked in the natural foods industry for most of his adult life but his long experience has neither lulled him into complaisance nor rendered him immune to inspiration. Perhaps it is his other life as a jazz singer in the style of Mel Torme—when he was little more than a toddler, he told his *nonna* that he wanted to be a song-and-dance man and he performs frequently at house concerts—that keeps him open to delight, a sensibility that shapes his store in subtle, inviting ways.

A busy shopper might not notice but a thoughtful customer surveying the shelves will spot Ernie's savory inspiration, a salt bar, tucked at the far end of an inner row.

Salts from around the world, some—that culinary workhorse Diamond Crystal Kosher Salt, for example—are familiar but most are new, even to dedicated salt aficionados. There are dozens of brands of sea salt, some ground as fine as powder, others in large crystals, a few ideal for a salt grinder.

The heart of the salt bar rests on a middle shelf, where nearly two dozen glass jars filled with salts that glisten like colorful jewels in the light beckon the curious and the eager. Among the selections are Kala Namak, traditional black salt from India; Murray River, pale peach flake salt from Australia; Hawaiian alaea salt, tinted brick red by natural clay; and Sel Gris de Guerande, the famous gray salt from the northwest coast of France. There are two "lava" salts, tinted black with activated charcoal.

Some of the most intriguing salts are those that have been smoked; a single grain dissolving on your tongue evokes delicious possibilities. Sprinkle a few crystals over deviled or scrambled eggs, cheese fondue and thinly sliced salmon or halibut gravlax.

All of the bulk salts may be sampled but above the jars that hold them is a row of small tins of flavored salts that must be taken home to be tasted. You'll find salts flavored with everything from rosemary, garlic, lemon, and lime to ginger, black truffles, wine, espresso, balsamic vinegar, and more, each a perfect hostess gift.

marshes of Brittany and tasted the salt that is sold, so casually and cheaply, alongside the road and at the outdoor markets of the area. But it was pricey (in some instances, well over ten dollars a pound) and uncommon in the American kitchen. In 1997, Holly Peterson Mondavi, a chef in the Napa Valley wine country in California, founded her company Sea Stars and began importing and packaging an elegant, coarse sel gris from Guérande in France. By 1998, sel gris was everywhere. The race was on. Now it is nearly impossible to name all the brands.

Just two hours from Paris (via the high-speed train), windswept Brittany has a wild, ancient feel. The region has a unique cuisine, shaped in part by its signature salts. Here in Brittany locals prefer salted butter, the only region in France where this is true. Lamb raised on these salt marshes is called pré-salé, that is, "next to the salt," and is praised for its succulence and flavor. And then there are Brittany's prized oysters, so enduring and abundant for centuries, like its salt, yet so fragile and threatened today.

Salt, Salt Everywhere

Artisan salt farming is not a French monopoly. In Portugal, a natural artesian well feeds the salt ponds of Rio Maior, an inland village, where for centuries families have raked and harvested their sun-dried salt using traditional methods. Here, houses are made entirely of wood—walls are joined by wooden dowels; even locks and keys are made of wood. Salt corrodes metal (if you live where it snows you know this; it was news to me, a child of the California sun), and does so quickly, yet it preserves wood for centuries.

Farmers are barefoot in Rio Maior, too (a practice that prevents them from accidentally kicking dirt and rocks into the delicate salt), but not bare-chested, bare-headed, or bare-handed. They wear traditional knit hats to protect themselves from sun and salt, and use hand-carried baskets to transport the salt from the crystallization ponds to nearby piles of fresh

salt. The salt—smallish crystals with a clean, briny flavor—is sold through local cooperatives.

The Mediterranean Sea is saltier than the Atlantic Ocean, as inland seas always are. The increased concentration shortens the time required for the water to evaporate and the salt to crystallize. The quieter tides make salt farming an easier task than it is along the Atlantic or Pacific coasts. Most Mediterranean salt, such as La Baleine, a popular French sea salt now owned by Morton's, the American company that made "when it rains it pours" a household slogan, is refined in modern facilities.

Trapani, on the northwestern tip of Sicily, is different. Salt has been farmed there since the time of the Phoenicians, and it seems as if little has changed. Ancient windmills are still the only power, save the sun. These enormous windmills—each is named and most are several centuries old—power the pumps that transport the seawater into the drying pans, and also turn the stone wheels that grind the salt crystals. No barefoot farmers here, though; the workers wear tall black rubber boots. The pure white salt with its angular crystals is extremely hard, dry, and slow to dissolve; it's a good salt to use in salt mills and for salting the water for pasta.

Snowflakes on the Windshield

Professional chefs differ when it comes to the salts they prefer. Some don't care (I've been in those kitchens, seen the tin shaker on a rack above the range, watched as a shower of table salt mixed with powdered black pepper rains onto virtually everything); some prefer what they perceive as the more complex flavor of sea salt, even when they aren't quite sure exactly what sea salt is (and even when it's the same product as that table salt).

A lot of chefs like kosher salt, by which they mean the coarse salt sold under the Diamond Crystal brand (now owned by Cargill Corporation), which is easy to grab between your fingers, the classic "pinch" of salt. There are other kosher salts on the market, made using different techniques, but Diamond Crystal is unique. It is my salt of choice for cooking, my default salt, the one I carry in my purse or pocket in a discreet little wooden box, just in case. It is different, and its qualities and the method by which it is produced are worth a look.

It was a continuing debate with my friend Daniel Patterson, an extraordinarily talented chef with a palate I admire, about our favorite salts (he prefers a French sea salt for general cooking) that set me on the trail of kosher salt's story. It wasn't long before I knocked on the door, electronically speaking, of the late Skip Niman, a chemist who began working for Diamond Crystal in 1959. He was, during his long career, probably the world's leading expert on the Alberger method of producing the coarse crystals we call kosher salt.

"It will be easiest for you to understand important differences between kosher salt and table salt if you look at the two under a magnifying glass," Skip suggested just a few minutes into our first conversation. I followed his instructions, and urge you to do the same. Place small separate piles on a dark surface—a black plate or a piece of construction paper is perfect. Look at each through something that will magnify it a little. The table salt looks like a pile of perfectly shaped ice cubes. Now; think of what happens when you drop ice cubes on a car windshield. They slide down, melting very slowly, leaving a trail of water in their wake. That's how granulated salt acts in food and on your tongue.

The Diamond Crystal Kosher Salt crystals are uneven and jagged, not unlike snowflakes. What happens when snowflakes fall onto a windshield? They melt instantly (don't argue about frozen windshields and icy buildup; you know what I mean), which is what kosher salt does—it dissolves in a burst of flavor. (If you have another brand of kosher salt around, take a look at that, too. The crystals may be solid and flat, as if run over by tiny steamrollers; or they may be a piles of little cubes, flattened and fused together.)

Most food-grade salt, including much sea salt, is dissolved (or redissolved, in the case of solar-evaporated salt) and dried in an enclosed vacuum pan; the result is granulated salt, those little cubes you saw so clearly

Flakes of Joy

Ahh, flake salt, those dry, fragile crystals, Skip Niman's snowflakes, how transformative they are. If all you have ever used is the tiny hard cube known as table salt, you are in for a surprise that will, if you pay attention, take your breath away. Because flake salt is dry and fragile, it dissolves quickly, causing flavors to blossom quickly on the palate. It should be your daily salt, your default salt, the one you can't live without. When I travel, I carry some in a little salt box tucked into my purse.

The most readily available flake is Diamond Crystal Kosher Salt, the largest of several flake salts grown in the same large circulator. The salts are screened for size and if you put the smallest ones under a powerful microscope, you'll see that they, too are tiny uneven flakes. This kosher salt is inexpensive, which makes it practical; it is delicious and infinitely useful but not dear. Toss it into pasta water, use it for brining, preserving and fermenting, for finishing.

Other flake salts are interesting, delicious, and fun to use as condiments. Maldon Sea Salt Flakes is produced in Essex, England, by heating a saturated brine, similar to how Diamond Crystal Kosher Salt is made. These crystals tend to be a bit larger and they crack like briny jewels between your teeth. Recently, the family owned and operated company added smoked salt to their classic white salt. These two salts add a delightful flourish to all manner of foods, from watermelon and pineapple to grilled lamb chops.

Other flake salts—Halen Môn from Wales, Murray River from Australia, Cyprus Flake Salt, Cyprus Black Lava Salt—are lovely, too, and appropriate for finishing dishes, not as default cooking salts. Because mouthfeel--the physical impact of shape and texture on the palate—is an essential aspect of these salts and because this quality vanishes the moment the salt dissolves, there is no advantage to using any of them in general cooking.

Care for a Sandwich?

Although the sandwich is utterly familiar to us as that famous dish consisting of two slices of bread with something wedged in between, the suffix "wich" has a long history in England indicating a village where salt was produced. Such places as Droitwich in Worcestershire and Northwich, Nantwich, and Middlewich, in Cheshire, all produced salt. The term seems originally to have indicated the group of buildings where salt was made, as well as the wich-houses, where brine was evaporated. The word "sandwich" as it applies to food is said to have been named for John Montagu, fourth Earl of Sandwich, who once spent twenty-four hours at a gambling table and ate nothing but roast beef between two slices of bread, something he could hold with one hand while he continued to play.

under the magnifying glass. The uneven crystals known technically as grainer and casually as kosher salt are made by heating saturated brines, thus hastening evaporation and crystallization, which occurs on the surface of an open pan. The salt grows downward in the brine, forming hollow pyramids with thin sides known as hopper crystals. Because of the cost of fuel, very little true grainer salt is produced today; Alberger salt, a patented technique developed by J. L. Alberger in 1889 and owned today by Diamond Crystal, has replaced it.

Alberger salt is made by dissolving and extracting underground deposits of salt. Sodium hydroxide and sodium carbonate added to the brine cause the most common impurities—calcium, magnesium, and sulfate—to precipitate out of the solution, leaving behind a semipurified brine, which is pumped through a series of pressurizers and heaters that increase the temperature to 290°F.

At this point, it is a rolling, boiling stew of salt, a supersaturated brine that is pumped into a long, cylindrical container called a graveler, filled with large lake stones. Baffling plates force the brine up and down through the graveler, and as it washes over the stones, calcium sulfate, already looking for solid ground, clings to the handy rock surfaces as the brine continues on to the next stage, a series of flashers in which the temperature and pressure are gradually reduced.

Chapels of Salt

Being sent to the salt mines is not, by any measure, a pleasant employment opportunity. The work was extraordinarily rigorous and dangerous before modern equipment made the process easier and safer, and both prisoners and slaves have been forced to extract rock salt from underground mines. When they were able—away from the watchful eyes of their bosses, or else in their favor—a few of the workers left their marks in carvings of salt along the many tunnels in the mines. These were often religious carvings, which provided a place of worship for workers who remained underground for some time; they also were thought of as guardians to protect the miners from disaster. Some of the most famous carvings are in the Royal Wieliczka Salt Mine near Krakow, Poland. Little chambers, small chapels, and one huge cathedral are filled with small and large religious statues and bas-relief tableaus, all carved out of the rich black salt of the mines.

In many of the underground rooms, elaborate chandeliers made of beads carved of white salt are suspended overhead and fitted with electric lights. As the mine has become increasingly popular as a tourist attraction, newer carvings have been added; some depict various Polish folk tales and myths, some historical figures such as Copernicus; others celebrate the long tradition of working in the mines, which have been in operation for over seven hundred years. There's a café, a small shopping mall that sells souvenirs of salt in various forms, and a dance hall that can be rented for concerts, weddings, and other private events. Rock salt is no longer extracted from this mine; today, all the salt mined here is extracted as brine and evaporated above ground in vacuum pans. So much salt has been removed over the centuries that to dig more out would risk weakening the ground and causing the town of Wieliczka to collapse into the deep caverns, something that has occurred in other parts of the salt-world, in Northwich, England, for instance.

In Wieliczka, a vast sea of salt is dissolved in underground salt lakes, which bloom a brilliant emerald green when they near saturation. Curved bridges and steep wooden staircases, underwater lights, and wave machines that move the water and throw rippling reflections on cavern walls have been added to enchant visitors even more.

Visitors pass through just a small portion of the vast web of tunnels and hallways here, moving in large groups scheduled one after another throughout the day. As I walked from the lobby and began the descent into the caverns, I looked at the black walls and wondered: *Is this all really salt?* I had to know, to really know, not take history's word for it. When no one was looking, I licked a finger, rubbed it along low on the wall for a few seconds and then put my finger in my mouth.

Yes. It is all really salt.

Amami-öshima island is a forty-five-minute plane trip off the coast near Tokyo in Japan, a country where the government still taxes salt and oversees its production. Here, salt is made by using a fast, modern process that concentrates the brine before pumping it into ponds for final crystallization. Darrell Corti, a grocer in Sacramento, California, describes Oshima Island Red Label salt as remarkably sweet and recommends it as a condiment salt. Although I always try to defer to Darrell's superior palate and greater experience, and even though I cower at the prospect of disagreeing with him publicly, I confess that I detect no sweetness at all in Oshima Island salt, and find its small fluffy crystals impractical in a salt cellar (salt spoons don't work, and it wedges itself under my fingernails) and impossible in either a mill or shaker.

The United States sits upon a huge salty empire, enormous underground deposits left by ancient seas that stretched across the Great Lakes

Himalayan Salt

As I wrote the first edition of this book, Himalayan salt was one of dozens of salts available from specialty retailers but it was not at the head of the pack; few if any purveyors were touting the nearly miraculous healing qualities attributed to it today, when advocates promise almost everything from this salt save immortality.

The salt, which hails from the Punjab Province of Pakistan, is mined from ancient inland deposits from the Salt Range, southwest of the Himalayans. It is halite, an unpurified rock salt with about 4 percent trace minerals, including iron oxide, which is what gives it its rosy hue.

The salt seems a bit exotic, given its pedigree. How much there must be in that Salt Range, how enchanting that place must be to give us such abundant beauty. *Carve me a bed of your rosy salt and let me drift into savory dreams.*

But will eschewing other salts change your life, return you to youthful energy, improve your love life, and make your cooking soar? As lovely as the salt is, it will not. The reason to use it is because you enjoy it but don't fall for the magical claims.

One claim is particularly misleading, that Himalayan salt has less sodium that other salts, an attribute not limited to this salt. Measure by measure, this and many other salts have less sodium because of their bulk density. There is not just less sodium in a teaspoon of this salt or hundreds of others when compared to, say, table salt or finely ground salt. There is less of everything because the uneven crystals have more space around them; table salt is more compact. The actual composition of the salt is unchanged; remember that ionic bond, so perfectly square, so immutable.

A telling demonstration of just how far our fascination with salt in general and Himalayan salt specifically has gone can be found in New York, where spas with both beds and rooms made of blocks of rosy-hued salt from inland caverns in Pakistan are increasingly popular. The pure air and and negative ions of these rooms are said to detoxify the

skin and lungs and facilitate relaxation, easy breathing, better sleep and general overall health. There is at least some validity to these claims, as rooms—and not just of Himalayan salt—in salt caverns, with their unique microclimates and stable atmospheres, have been used for centuries for convalescence, especially from respiratory illnesses.

region, the Midwest, the Gulf, the desert. In Utah there's the Great Salt Lake, which we all know, and in Louisiana, Avery Island, a better-kept secret that is home not only to massive salt mines but also to Tabasco sauce, one yummy place indeed.

In 1887 Ben Blanchard accidentally discovered that Kansas was sitting on top of a vein of salt that turned out to be about 30 miles wide, 150 miles long, and 325 feet thick. He had been drilling for oil. (Usually it's the other way around, you find salt first, in the form of a natural salt dome, and where you find one there's a good chance there's oil.) At the height of the industry that Blanchard launched, there were eighteen salt producers in Hutchinson, but most were bought out by Morton Salt Company, the first company in the United States to package table salt.

So much salt; salt everywhere, too much salt to tell you everything, too much salt even to know. From the high plains of Peru and the salt lakes of the Himalayas to the beaches of Bali and Baja; from ancient salt mines in Africa to the deep caverns of Germany, Austria, and Poland, there is so much salt, more than we will ever, could ever, use. It's heartening to know, isn't it, that something we need so desperately, so absolutely, is ours in such abundance?

Red Rocks, Crimson Crystals

A saturated brine pond turns red because the protoplasm of *Dunaliella* algae develops a red pigment as salinity increases; tiny brine shrimp add

A Gift of Aloha: *I komo ka mea 'ai i ka pa'akai*

When I was a child in Hawai'i, my Kūpuna drove my brother and me around the island almost every Saturday. We stopped at their choicest beaches and wading ponds to pick limu, to catch black crabs on the rocks as the sea crashed upon us, and to dive for fish, which we enjoyed for dinner. I remember watching my grandparents clean and prepare the mea 'ai in which pa'akai, or sea salt, was the main ingredient. Without pa'akai, the food had no character and, therefore, did not fully express the intention and desire of the cook to prepare a festive, delicious, and nourishing meal for their loved ones.

The Indigenous Native Hawaiians, or Kānaka 'Ōiwi, lived by many principles that aligned them to their natural world. One of these principals is *Mālama i ke kai*, take care of the sea. Kānaka 'Ōiwi of yesterday and today look to the sea for sustenance and for cleansing of the inner as well as the outer self. Pa'akai was used not only in food for flavoring, minerals, and preservation, but, just as importantly, for ritual healing and cleansing ceremonies. Oli, or chants, were done for blessings and cleansing of homes, canoes, heiau (temples), people, and more, with either fresh water mixed with pa'akai or with kai, seawater. The pa'akai was the most important ingredient in administering the healing/cleansing aspect of a ceremony.

Among the many roles that pa'akai play in the Hawaiian Culture, I feel that they all contain an underlying intention and purpose, to call us back to our connection with the sea and realign us to our natural state of whole health.

I offer these words as a gift of Aloha, "I komo ka mea 'ai i ka pa'akai." It is the salt that makes the food more welcoming,

—Kumu Hula Shawna
Kealameleoku'uleialoha Alapa'i,
Halua Hula Nā Pua O Ka La'akea.

an aura of orange. Rock salt (from Utah and coastal France, to name two sources) is studded with trace amounts of reddish-tinged iron. In Hawaii, there's a pale orange salt that occurs naturally along the shores of both Kauai and Molokai; red clay, or alae, is responsible for the color. For centuries a deep reddish brown salt, mixed naturally with dirt, sand, and iron-rich clay, was gathered for both daily and ceremonial use. Today, some red clay is already present in the evaporation ponds of Kauai, but most is introduced to create Hawaii's signature salt. The bond is weak; dissolve a spoonful in a glass of water, and a band of red clay settles to the bottom while the sodium chloride vanishes into solution. Hang a thread in the water and pure white salt crystals migrate up it, a chemical expression of *two's-campany-three's-a-crowd*.

Native Hawaiians use their salt in ceremonies and rituals, such as the blessing of the canoes, when a few grains are sprinkled on the tongue. Red salt is also used in poke (cubed raw fish) and to finish cooked fish dishes.

"It is special because it comes from the other islands, because it was once so hard to get," explains Jim Medeiros who lives on the Hawaii Moke and whose great-great-grandmother was one of King Kamehameha's mistresses. "It isn't used as everyday salt."

Gone with the Wind, but Saved in the Salt

Underground salt caverns are valued for their unwavering temperature and humidity. Underground Vaults and Storage, a rental facility in Hutchinson, Kansas, leases hundreds of storage units in its underground facility housed in an old salt mine. Thousands of irreplaceable cultural treasures are stored here, such as the original prints of thousands of films, including, it is rumored, *Gone with the Wind* and *The Wizard of Oz*. Too bad there isn't a theater, too.

The formula for classic Coca-Cola is also reported to be kept in one of the salty vaults.

Some do use it daily, saying that the minerals in it are good for you, and taste good, too. Why would you use any other salt? they wonder. Unlike many specialty salts, Hawaiian alae salt is inexpensive, selling in supermarkets in Hawaii for about two dollars a pound, slightly more in California. But will the low prices last as Hawaiian salt is discovered? Probably not. In mid-1998, high-end markets and catalogues were adding it to their product lines, at considerably higher prices. Before long, there were a dozen or more Hawaiian salts, some actually from Hawaii, others name-only endeavors manufactured far from the islands. Hawaiian black salts, one named after Kilauea, are beautiful but not traditional. They are manufactured salts, combined with activated charcoal, which can tint them as black as dried lava. These salts played no role in traditional Hawaiian culture but they have a certain sexy appeal.

Salting the Ritz

Ed Walsh, a chef in northern California, comes from a long line of restaurant workers. His parents met at the Hotel Ritz in Paris, where they fell in love and ran off together. As a young man, Ed went to Paris, too, where he worked in the dining-room kitchen (there were other kitchens for room service) of the Ritz Hotel in the late 1960s. Every night, Ed remembers, the floors of the kitchen were covered with a thick layer of rock salt to keep out rodents and cockroaches. Every morning the salt was swept up and reused that night.

Jewish Barbecue, the Best Bacon, and Other Salty Miracles

The mysterious ability of salt to affect flavor beyond adding its own character may be best revealed in dry-salting and brining. Short-term brining adds flavor to bland foods and succulence to normally dry cuts of meat. Long-term brining transforms both taste and texture, and preserves foods as well. Dry-salting intensifies natural flavors, contributes new ones, and preserves; it is often used on foods that will be smoked.

Brines and dry-salt rubs are used extensively in commercial meat curing. Both brines and salt rubs penetrate meat at the rate of about an inch every seven days, but today brine is commonly injected directly into meat to mimic, however ineffectively, the slower process of making bacon, ham, corned beef, pastrami, and the other products that rely on salt curing for their characteristic flavor and texture. Authenticity and quality are both compromised by this faster method, but it has become so common that many people have never tasted the real thing. You won't find much of the real stuff in major markets, but there are plenty of mail-order sources for meats cured the old-fashioned way. It is easier to find authentic dry-salted smoked bacon than it is to find true country ham. Many of the producers are in Virginia and Kentucky.

When a box of bacon, hog jowls, and ham arrives from R. M. Felts Packing Company of Virginia, it's like my birthday, which conveniently comes in the summer, just about the time the first backyard tomatoes ripen. The first BLT of the year is always a sacred ritual, a lusty Bacchanalia, and it took a leap heavenward when I discovered Felts's bacon in The Art of Eating, Ed Behr's newsletter. I agreed with Ed; at that time, I'd never tasted better bacon.

Felts's bacon is pork and salt, nothing more. The rind is intact, and the bacon comes attached to the ribs or removed; it's a bit of a trick to cut it off the rib yourself, but there is very tasty, lean meat tucked between the ribs that is lost when the ribs are cut off mechanically. (Bacon lovers will get the hang of it in no time.) The bacon produced at this Virginia farm, operating since the 1930s, is rubbed with salt and left for not quite two weeks, during which time the salt penetrates right through the meat, which is then washed, smoked for three days, and air-dried in a warm room for another week. *Et voilà!* it is bacon. Before it is sold, Felts's bacon is rubbed with crushed black pepper.

About the time the first edition of this book was published, the artisan bacon movement began to pick up steam and in a few years would

become a national pastime, an obsession. Black Pig Meat Co.'s bacon, produced by Chef John Stewart, a friend and colleague in Sonoma County, became a national success. John and his wife, Duskie Estes, have won a national competition that crowned them King and Queen of Pork. I was a judge and Francis Ford Coppola, also a judge, and I sat across the table from each other as we savored a dark chocolate–bacon brittle lollipop that very likely secured their crown.

Today, farmers markets have become among the best sources for artisan bacon, with small ranches offering a delicious range of styles and flavors.

Felts's principal product is dry-salted smoked ham, intensely flavored meat that bears no resemblance to the watery ham sold in supermarkets. It is as close a thing as we have to the prosciutto of Italy and the jamón serrano of Spain, and it is probably best to think of it in this way, rather than as most of us think of domestic ham. A big slice with red-eye gravy and mashed potatoes is likely to be too intense, too salty, for all but the most hard-core devotees. Boiling the ham leaches out much of the salt, but I think it's best in small portions, served raw as an appetizer or used to flavor cooked greens, beans, and stews. Scalloped potatoes flavored with this ham are sensational.

"Pastrami is Jewish barbecue," John Harris, the author of *The Deli Book* tells me, "and what makes it good is salt, pepper, fat, garlic, and spices." The four basic food groups, some might say, plus spices.

Pastrami refers to a process rather than to a specific product. It can be made with duck, turkey, lamb, venison, and wild boar, though classic New York pastrami is made using a tough piece of meat near the belly of a cow known as the plate, or navel. It is a cheap cut, tough and fatty, that needs long, slow processing to tenderize it and leach out some of the fat. When you bite into a pastrami sandwich at, say, Katz's Deli, you are tasting, in part, the alchemy of salt, the magic of pepper.

Both corned beef and pastrami begin with pickling—or corning, after the large salt crystals called "corn" in England—in a salty brine. Corned beef is brisket, a less fatty piece of meat closer to the chest of the animal, that, after pickling, is ready for sale, though it must be boiled for several hours before being eaten. Pastrami goes through further processing, beginning with a spice rub. Some manufacturers use lots of garlic, others cloves and coriander; some emphasize black pepper. Next, the pastrami is hot smoked; it is fully cooked when you buy it, but tastes best after steaming, which further tenderizes it and transforms the fat into a delicious, juicy miracle.

Of course, it is not only flesh that may be preserved with and in salt. Caviar is between 3 and 4 percent salt, and without a proper and careful salting, it's … well, it's just fish eggs. Sauerkraut is simply (and miraculously) cabbage fermented in salt. Cucumbers become pickles in a salty brine, as do a host of other vegetables. Without salt, there would be no Korean kimchi. Although some olives are made without salt, most rely on salt either to leach out their bitterness or to add flavor when something else has done so.

Salt is crucial in bread baking; it controls the activity of the yeast, slowing it down so that the flavors of the flour have time to develop. Salt also firms the texture of dough, tightening the gluten. Many French bakers and an increasing number in this country swear by sel gris, but not all bakers agree that there are differences in the way salt influences the taste of bread. Craig Ponsford of Artisan Bakers in Sonoma, the baker who won the 1996 World Cup of Baking for the U.S. team and today operates Ponsford's Place, a bakery and experimental kitchen in San Rafael, California, likes both fine sea salt and kosher salt because they dissolve quickly; subtle differences detected between salts on the tongue, he says, do not create different tastes in bread. I tend to agree. Some (but not exclusively, as Ponsford's scrumptuous breads clearly demonstrate) of the bakers who use sel gris, indeed, make superior bread, but I've not been able to identify salt as the most crucial variable. Rather, it seems to be an overall dedication to excellent ingredients and a special care taken with all of the steps in the process that give superior results.

Cheesemaking the world around would be vastly different were there no salt, which flavors curds, regulates aging, and inhibits or stops the growth of bacteria. Cheeses may be massaged with dry salt or washed, sometimes daily, in a salty brine.

Feta, the traditional sheep's milk cheese from Greece and Bulgaria (France and the United States also produce feta, often using goat's or cow's milk), is probably the best known of the brined, or pickled, cheeses. For classic feta, fresh curds of unpasteurized sheep's milk are ladled into molds, packed, drained, and returned to the whey after it is seasoned with salt. Two weeks later, the cheese has a pleasingly salty taste.

In Poland's Podhale Valley, a region known as the Highlands, sheep's milk and salt are combined to make a delicious fresh cheese called Bryndza; it is creamy in texture, almost like a fresh ricotta. Another sheep's milk cheese, Oscypek, is pressed into molds, most of them cylindrical but some whimsically shaped into hearts, animals, and toy drums. Some of this cheese is sold merely aged, but most of it is lightly smoked first. The smoked cheese is frequently sliced, fried, and served as an appetizer.

Band of Black Gold

If salt is essential in cheesemaking, peppercorns, though not crucial, add a savory flourish, a definitive character, to certain regional cheeses. Italian pepato (also made in Latin America and the United States), a sheep's milk cheese just a few peppercorns away from pecorino, is studded with both cracked and whole peppercorns. Sicilian pecorino has a band of black peppercorns running through its middle. Several Greek cheeses include peppercorns, such as Métsovo from the village of that name in the Pindus mountains. Kopanistí, a blue-veined cheese, is kneaded with salt after the mold has grown but before the cheese is aged; as it ripens, it develops a mild peppery flavor, the result of interaction between the salt, cheese, and mold, rather than of introduced pepper. One of the classic new American goat cheeses, produced first by Laura Chenel and now by countless small and large cheesemakers, is a five-ounce log of chabis coated with crushed black pepper.

Sicily's remarkable ricotta is a pure expression of salt's alchemy. If all you know of ricotta is the cloying stuff in American supermarkets, you may by wide-eyed and suspicious at the idea of ricotta being good, let alone remarkable. But in Sicily, fresh ricotta, its gently pressed curds still as distinct as veins in marble, is one of the finest fresh cheeses in the world. It is both sweet from the good milk with which it is made and salty. Enjoyed while still warm from a ceramic bowl at a farmhouse in Sicily on a warm morning, the scent of citrus blossoms in the briny sea breeze is as good as breakfast gets. And the cannoli made with this ricotta? It is insanely, ridiculously, exuberantly over-the-moon delicious. I made the mistake of asking to share one with a dinner companion because I was, or thought I was, full. He offered me the first bite, I closed my eyes and when I opened them, the cannoli had been replaced by a satisfied glow on my friend's face.

Italy's Taleggio and similar cheeses, such as California's teleme, are washed in brine and sometimes briefly submerged in brine, which gives the cheese a thin, pliable rind. Rubbing cheeses with dry salt encourages the formation of a harder, tougher rind, Parmigiano-Reggiano being the classic example. In many cheeses, salt is mixed directly into fresh curds; this is the case with Cheddar, Monterey Jack, and most goat cheeses. Some cheeses, such as Emmentaler, are soaked in a brine before being sent to the aging room.

Each of these techniques of introducing salt to the cheesemaking process is essential. Attempts to make salt-free cheeses have been largely

Forty Salts & Five Thousand Years

Nearly five thousand years ago, what is perhaps the world's first treatise on pharmacology, the Peng-tzao-kan-mu, was published in China. The ancient text includes a significant section on salt, in which forty different varieties are explored. The salt production techniques described are said to be strikingly similar to those in use today.

—Salt Institute, Facts About Salt

unsuccessful (as have attempts to make low-fat, nonfat and dairy-free "cheese," but that's another story); the cheeses lack character, distinction, and taste, which cannot be created by adding salt to cheese at the table, when it is too late for it to work its transformative magic.

Salt preserves food, among other things, by drawing out moisture not only from the material to be preserved but from the cells of the bacteria that, left to their own briny devices, would hasten spoilage. Sausages, especially those made of fine emulsions such as bockwurst and bratwurst, also rely on salt to help maintain an even distribution of ingredients. Without salt, there would be no andouille, pepperoni, chorizo, linguica, or kielbasa, no knockwurst or liverwurst. There would be no turkey jerky for hikers to stuff into their backpacks.

Salt can be a sneaky conspirator, too; hollow crystals readily absorb and hide the fat in potato chips, tortilla chips, and other prepared foods that have an outer layer of salt. "Greaseless," we are told, and it's true, though deceptive. The fat is still there.

The Lonely Life of a Nevada Camel

Salt was essential in the mining and refining of silver, discovered in Nevada in 1859. For several years, the nearest salt was harvested from the marshes of San Francisco Bay. With the railroad still a decade in the future, horses were used for the arduous journey until Otto Esche had the bright idea of bringing in camels to do the work. Fewer than half of the thirty-two Mongolian camels he imported survived. Although the camels eventually adjusted to the western desert, they made the other animals nervous and led to one of our more bizarre laws, enacted by Virginia City, that limited the use of camels to the hours between midnight and dawn. The discovery of salt closer to the silver mines made the long trek unnecessary. The unfortunate camels were abandoned in the Nevada desert and perished.

—Nilda Rego, TimeOut, 21 April 1996

Pepper, too, is a principal ingredient in prepared and preserved foods. Spices were once primary ingredients in preserving foods (spices kill nearly all food-borne bacteria, hence their popularity before the age of refrigeration), but today most spices, including pepper, are used primarily for the flavor and aroma they contribute. The food packaging industry is the major consumer of pepper in this country. In fine-ground, light-colored sausages such as bratwurst, ground white pepper is used so that the meat will not be speckled with black. Pastrami, bacon, gravlax, and other cured meats and fish often have an outer layer of cracked black peppercorns.

Koshering and Kashering

When it comes to meat, the process of kashering—preparing foods according to Jewish dietary laws and traditions, that is, making foods kosher—includes the removal of any blood that lingers in the flesh of

The Pork Barrel

When a politician hands out the pork, what exactly is being given away? Salt pork. For the pioneers traveling west, it was one of just a few essential staples. Europe had bread; young America had pork belly preserved in a barrelful of salty brine. Salt pork sustained soldiers during the Civil War, and when Northern troops destroyed a new salt mine on Avery Island in Louisiana, Confederate soldiers faced a debilitating salt famine. Lack of salt, and salt pork, is said to have won the war.

Salt pork is used today—in collard greens braised with salt pork, for example—but it's no longer ubiquitous. Faith Willinger, an expert on Italian foods and cooking, considers salt pork better than the pancetta available in the United States. Rub it with cracked peppercorns, she advises, wrap it in plastic, and use it in recipes that call for pancetta. I find the pancetta in my local Italian market just fine, but I use salt pork, too, peppered, just as Faith says. When it comes to preserved meats, the more the merrier.

A Taste of Sin

In France, reports Waverley Root in his remarkable book *Food*, the profession of "sin eater" flourished for a time. When a family member died, a saucer of salt and a piece of bread were placed on the chest of the deceased before a practitioner of this singular trade was summoned. The sin eater arrived, ate the bread and salt, and thus, it was believed, the sins of the departed, saving the loved one from a long stay in purgatory or even eternal damnation. The flaw, at least in this world, was that sin eaters were despised and the profession died out for want of practitioners. This symbolic function, that of salt absorbing unwanted material, has a correspondence in the real world—salt draws out and absorbs liquids and soluble impurities, making it effective in the process of koshering, preserving, embalming and cleaning. Spill a glass of delicious pinot noir on your white tablecloth? *Quick! Grab the salt, pour a thick layer, and watch as it turns red.*

the animal. Kashering begins with meat from an animal slaughtered by an approved method, though this does not, of course, involve our topic, salt. Salt is a key ingredient in a step further along in the process and its importance has given us the common term "kosher salt," which Rabbi Jonathan Slater of Santa Rosa, California, explains is a bit of a misnomer. Kosher salt is merely coarse salt: it is more accurately called "koshering salt."

Kashering is sometimes done at home, and sometimes at the butcher shop before the meat is purchased. For example, kosher ground beef will have been salted and soaked before being ground; delicatessen meats begin with kashered meat, as well, and no further processing is necessary once they have been purchased. A steak, a roast, or a leg of lamb, however, likely will not have been kashered. If a piece of meat is to be broiled or roasted, it does not need to be salted; the blood will

Sacred Salt, Psychic Pepper

If you need peace and quiet in a noisy house, grind together some salt, sugar, and Dragon's Blood palm, place it in a closed container, and hide it in the house. If there is negativity or a lack of harmony in the home, soak a branch of rue in saltwater and shake it throughout the troubled area. To protect your home from being struck by lightning, toss some salt into a fire. Add pepper (a masculine plant, ruled by Mars) to amulets to protect yourself against the evil eye (and your own envious thoughts). Mix pepper with salt and scatter the mixture outside to dispel evil. If you're troubled by ghosts, carry salt in your pocket to render them invisible. If you have an exorcism scheduled, don't forget the pepper, an essential ingredient in the ancient ritual of casting out demons.

—Cunningham's Encyclopedia of Magical Herbs

drain away as it cooks. Meat for stews, soups, and braises may need to have the blood removed before cooking, which is achieved by soaking the meat in water, draining it, and then covering it with a layer of salt and letting it sit for thirty minutes or so, during which time a coarse salt, rather than dissolving as granulated salt does, draws out blood and other impurities. The meat is then soaked in cool water for a few minutes and dried; it is then ready to be cooked. Poultry must undergo a similar process. If you keep a kosher kitchen, you don't need me to tell you what is done. But if, like me, you've only heard of kosher kitchens and have been intrigued by the practices, perhaps I've clarified the mystery. Kosher salt is not like holy water—it is not blessed by a rabbi and it is not imbued with magical powers. Rather, it performs a specific, tangible function.

Tiny seed crystals of salt form instantly as the brine is pumped from the last flasher into an open pan shaped like a huge figure eight with the

Versatile Salt

Salt has more than fourteen thousand uses; less than 4 percent of all salt produced each year goes into food. It is the second most important element—sulfur is the first—in the chemical industry, and is used in the manufacture of fabrics, glass, cosmetics, and ammunition. It is crucial to agriculture and has dozens of uses in the home that have nothing to do with eating, or at least not directly. You can polish copper by rubbing it vigorously with salt and lemon juice; you can extinguish a grease fire by covering it with salt. Saltwater poured down a drain will sweeten it. Salt is probably the most effective method to reduce the damage done by red wine spilled on a rug, tablecloth, or other fabric: Cover it immediately with a thick layer of salt (kosher absorbs liquid the fastest), which will draw out the wine and reduce the chance of a permanent stain. In the oven, sprinkle salt on food as soon as it spills: cleaning up after the oven has cooled will be much easier.

—The Salt Institute, Facts About Salt

middle part gone. Hopper-shaped crystals begin to form on the salt seeds and sink as they grow heavy. Paddles keep the brine circulating and when the salt has completely precipitated, the slurry—that is, the crystals and the remaining liquor, called bittern—is pumped into a centrifuge (picture a large washing machine). The liquid is spun off, the bottom of the centrifugal basket drops out, and the salt cascades onto a conveyor belt that takes it through a dryer. The dry salt is screened and graded into sizes, the coarsest of which becomes kosher salt.

Alberger salt has a lighter bulk density than granulated salt has: Equal amounts by volume are not equal amounts by weight. Simply put, this means that there is less salt in a teaspoon of kosher salt than there is in a teaspoon of granulated salt. Qualities such as bulk density, rate of solubility,

Deer Season in Michigan, Salt in St. Clair

Early one November morning in 2002, a couple of staff from Diamond Crystal Kosher Salt picked me up at my hotel in Detroit to make the hour's drive north to St. Clair, where the salt is made not far from Lake Huron. When we stopped for tea and gas, I noticed twenty-pound sacks filled with enormous red beets and huge carrots. They seemed to be everywhere, stacked next to gas pumps, near cash registers, near the doors of stores and markets.

As I paid for my tea, I asked the clerk, a young woman with long blonde hair and bright blue eyes, about the vegetables, as I had no idea whatsoever how they might be used.

"Deer season," she shrugged without looking at me, "sweetens 'em up."

The adventure continued and before long I was being lead through floor after floor of the production facility, where employees were clad in white from head to toe. Thick layers of salt covered everything, the walls, the floor, the stairs, the water pipes. No inch of wood or other material showed through the whiteness until we got to the top floor, where an enormous pan, a foot or so deep and shaped like a giant figure eight, glistened in eerie pale green light. A paddle slowly circulated in the soupy brine, moving salt crystals to a shoot on the far side of the pan, where they headed through a dryer and then made their journey down the floors to be screened by size, bagged, labeled and in some cases, as I noticed in a corner of a downstairs storage room, to be approved as officially kosher by rabbis charged with certification.

I stood there mesmerized by the process for a long time, and, if I remember correctly, asked for a chair and to be left alone for a bit. After askance looks from my hosts, they accommodated me and I sat there, transfixed as little crystals formed on top of the saturated brine, grew a bit and then, pulled by gravity, dropped down on top of the other crystals that had not yet been swept away by the paddle. Finally, propriety and etiquette required that I leave, though part of me wanted linger indefinitely. *Yes, I thought, I have stars—little stars, made of salt—in my eyes but I'm not crazy, honest. You won't have to call the authorities.* As we set out on the walk down six flights of stairs, I looked back at the brine and the slowly circulating paddle and knew it would resonate in my mind's eye for the rest of my time on earth. Soon, we were enjoying lunch at a lovely restaurant within view of the lake, where I enjoyed the most delicious fish—little lake perch, light breaded and deep fried—that I have ever tasted.

and blendability (dry ingredients mixed with Alberger salt will stay in suspension in a uniform blend; granulated salt will separate out and fall to the bottom, or, in an oily blend, leave striations) are important to food producers, but of little concern to home cooks.

What you need to know is that the saltiness of a specific quantity of salt will vary depending on the type of salt. If you taste as you cook and use your fingers instead of a measuring spoon, you will get the hang of perfect salting quickly. (If you're the curious type, you might enjoy knowing that a pound of granulated salt has 5,370,000 crystals and a pound of coarse Alberger salt a mere 1,370,000, or that it takes 9 minutes to dissolve 1,000 grams of granulated salt, 3.6 minutes to dissolve the same quantity of coarse flakes.)

Brining for Flavor

In the late 1990s, brining meat and poultry was all the rage. Chef and cookbook author Bruce Aidells, who is known among his friends as The Sausage King, is one of many advocates of brining meats to add flavor. It is particularly effective in adding succulence and juiciness to today's leaner meats, Bruce explains, especially pork; anyone who has gnawed on a rock-hard pork chop will appreciate the technique. When Bruce and his wife, the renowned chef Nancy Oakes, use brine for flavors, they use much less salt than is used in brines for preserving; they add other ingredients, too, including black pepper and other spices, vinegar, maple syrup, mustard, and, in one of Nancy's signature savory recipes, vanilla.

Shortly after this book's publication, I did twenty-four television segments for stations all over the country about brining turkey for Thanksgiving. The book's publicist pitched a story to *Martha Stewart Magazine* and apparently stunned Ms. Stewart herself, who responded that at her staff meeting they came up with thirty-eight ways to prepare turkey but brining was not among them. I was scheduled to appear on her show to demonstrate the technique but not long before my scheduled flight, we were informed that Martha decided she would rather work with a male chef for the segment. And that was that. *Oh, Martha, no salty kiss for you.*

Ed Walsh, former executive chef at Kendall-Jackson Winery in Sonoma County, soaks all meat (except rabbit, which doesn't have enough fat) and poultry in brine before cooking. Brining removes the adrenaline that is released into an animal's tissues when it is killed, he explains, and creates a much cleaner flavor. He begins with one cup of salt (the same proportion as Bruce and Nancy use) to each gallon of ice water and may add juniper berries, peppercorns, onions, and parsley. He brines pork for three days, and chicken for an hour and a half; for the latter he changes the brine every thirty minutes. Brining also improves the flavor of goose, duck, squab, and most game, he adds.

Ed grew up watching his parents salt almost everything that arrived in the kitchen of their restaurant on Long Island. Dried pasta was washed in brine to remove excess flour; produce was rinsed to extend its life; even seafood was washed in a saltwater bath. It was from Ed that I first heard the story of seppuku, a method of suicide among poor people and women in Japan. Traditionally, poor people did not have knives and women were not allowed to touch them. Instead of committing hara-kiri, a person would eat a pound of salt and die within a day.

Rejuvenating Salt

Salt invigorates our bodies as well as our taste buds. Add several handfuls of sea salt to a bath (or simply soak tired feet) and you'll feel revived. Several cosmetic companies, as well as the Grain & Salt Society (see resources, page 395), sell non-food-grade sea salt, some already mixed with aromatic oils and fragrances, for this purpose. Salt rubs and salt facials are even more rejuvenating—sea salt moistened with a little oil (almond, jojoba, or olive oil is best) and rubbed vigorously over the skin with a loofah leaves your skin feeling as soft and dewy as a newborn's. An exfoliant, salt removes dead skin cells, revives tired ones, and stimulates circulation. Some say that the minerals in French sea salts draw out toxins. For an exfoliating facial, mix equal parts of sea salt (crushed, not large crystals, which can cut) and olive oil, massage gently in upward strokes from the neck to the forehead, and leave on for ten to fifteen minutes. Rinse thoroughly with cool water and then apply a moisturizer.

I am no longer a fan of brining meats, a position that is fairly controversial, as many home cooks and professional chefs swear by the technique. But I don't think there is deep disagreement. In the late 1990s, good grass-fed meats and heritage poultry were still on the horizon. There were few alternatives to low-fat commodity pork and broad-breasted turkeys and they were not widely available and, in many cases, not very good. That has changed. Grass-fed meat, pastured pork, heritage pig breeds, and heritage turkeys are readily available almost everywhere. These meats have so much natural flavor and succulence that it is neither necessary nor advisable to introduce more liquid, which serves to dilute the flavors. When you learn to handle these meats well--they require different cooking techniques and times--there is little risk of drying them out and their concentrated flavors are the reason so many people are willing to pay the high prices they command.

A Fistful of Salt

Mahatma Gandhi captured the attention of the world and invented modern civil disobedience when he gathered a fistful of salt, a commodity that was controlled and heavily taxed by the British government.

"In India's hot climate, [salt] was an essential ingredient in every man's diet. It lay in great white sheets along the shorelines…Its manufacture and sale, however, was the exclusive monopoly of the state, which built a tax into its selling price…. for a poor peasant it represented, each year, two weeks' income.

"On March 12, 1930… Gandhi marched out of his ashram at the head of a cortege of seventy-eight disciples and headed for the sea, 240 miles away…The weird, almost Chaplinesque image of a little old half-naked man clutching a bamboo pole, marching down to the sea to challenge the British Empire, dominated the newsreels and press of the world day after day.

"On April 5, at six o'clock in the evening, Gandhi and his party finally reached the Indian Ocean near the town of Dandi. At dawn …the group marched into the sea for a ritual bath. Then Gandhi waded ashore and, before thousands of spectators, reached down to scoop up a piece of caked salt…. He held his fist to the crowd, then opened it to expose in his palms the white crystals, the forbidden gift of the sea, the newest symbol in the struggle for Indian independence.

"Within a week all India was in turmoil. All over the continent Gandhi's followers began to collect and distribute salt…. The British replied with the most massive roundup in Indian history, sweeping people to jail by the thousands…. Before returning to the confines of Yeravda prison, [Gandhi] managed to send a last message to his followers.

"'The honor of India,' he said, 'has been symbolized by a fistful of salt in the hand of a man of nonviolence. The fist which held the salt may be broken, but it will not yield up its salt.'"

—Larry Collins and Dominique Lapierre
Freedom at Midnight

The Saltmen of Tibet

For thousands of years, a few nomads have made an annual journey to a Himalayan lake in northern Tibet, where they rake and gather salt, pack it into sacks, sew them shut, and load them onto the yaks (nearly two hundred of the animals) that accompany them. Ulrike Koch tells their story in a two-hour film, *The Saltmen of Tibet*, that was released in 1998. The documentary was taped surreptitiously when the Chinese government denied permission to film. (The nomads had their own restrictions; Koch, a woman, was not allowed on the final days of the journey; her male assistants went on without her.) The nomads sing and chant, communicate in a secret salt language, and mold small animal salt-sculptures in the salt lake before they begin their return journey. The somber, elegiac tone of the movie is under-scored by the presence of ramshackle trucks—the first commercial harvesters of the sacred salt—at the lake when the nomads arrive.

PART II
Pepper

The King of Spice

Salt is indispensable. The earth is cloaked in salt and a salty tide moves within us, every moment lapping at the edges of our cells. Without it, we die.

Pepper is superfluous. Salt's culinary spouse is expendable, entirely unnecessary for existence (the ideal marriage, one partner frivolous, one keeping the house afloat). It is a gift, a luxury that we have come to take utterly for granted. "Saltandpepper" is nearly a single word in our culinary lexicon and the thought of doing without the pepper part is all but inconceivable. But should something occur to put pepper absolutely out of reach, we could live. We would mourn its loss, and with every bite be aware of its absence. But we would live.

In the end, we might get over the loss of pepper without too lengthy a struggle. We have no innate physical longing for it, as we do for salt. We delight in the presence of pepper in our food, but the memory of its taste would fade in a single generation; we cannot pass sensory knowledge to our children, we can only entice them with descriptions, always inadequate when it comes to taste, completely ineffective when it comes to smell. No one would believe our stories. If pepper vanished, it soon would become myth, an enchanting mystery, a tale of the King of Spices who one day walked off into the wilderness and vanished, the Atlantis of spice, nothing more than a tantalizing rumor. *Sure,* they would think, *and what is the moral of this tale?* The King soon would be replaced by . . . all-spice? long pepper? coriander? All have been used as we now use pepper—that is, in almost everything—at one time or another.

But don't worry about pepper vanishing—there is plenty, millions of vines of it wrapping themselves around the earth, a savory circumference, an equator of savor and taste and heat, a lusty belt around our middle. Pepper—green, black, and white peppercorns—is the fruit of a perennial vine, *Piper nigrum,* that thrives within fifteen degrees of the equator. It flourishes in tropical heat and monsoon rains. Native to the Malabar coast of southern India, pepper vines were introduced into Southeast Asia hundreds of years before the birth of Christ, and much more recently to Brazil, Micronesia, Madagascar, and Nigeria.

Although pepper vines can grow to thirty feet or more, most commercial plants are kept at between nine and twelve feet to make harvesting easier. Some farms keep vines to six feet to make ladders unnecessary. Each vine is trained around a central pole, or in some areas, a tree whose leaves shield the pepper from too much direct sunlight and whose roots fix nitrogen in the soil. A pepper farm resembles a Christmas tree farm, vines winding skyward like fast-growing fir trees.

Peppercorn berries grow in small clusters up to about three inches long; they resemble diminutive bunches of Cabernet Sauvignon grapes, loose and narrow rather than tight and plump. There may be fifty, sixty, eighty, one hundred berries per cluster, and the clusters do not all ripen at the same time. A single vine must be picked many times in a season to harvest all its fruit; all pepper everywhere is picked by hand.

Black pepper is the most common form of pepper, and for two reasons. It is the easiest to produce, and it has the fullest range of flavor and aroma.

Green peppercorns are harvested several weeks before black peppers, an advantage if you're worried about pests or disease, when the berry is still soft throughout, its hard core not yet formed. But the flavors are not fully developed at this point, either, and it is hard to keep the young berries green as they dry. Some are pickled in vinegar and some preserved in a salty brine, both wonderfully flavorful in many recipes but not exactly versatile as a spice. Until recently, freeze-drying was the only other method of preserving

Pepper in the Wild

Two varieties of pepper, *Piper vestitum* and *P. grande*, are found in the wild. Both bear small clusters that grow upward rather than down as *P. nigrum* clusters do. The berries of *vestitum* are light golden brown and black; *grande* berries are red. Both are hot but lack the complexity of cultivated pepper.

green peppercorns, but new techniques make it possible to air-dry green berries. To prevent them from turning black (for then they would merely be black pepper without its full range of flavors), they are soaked in vinegar or brine first. Green peppercorns have a mildly tart, fresh taste, full of the characteristic heat of black pepper, but lacking in its complexity and depth. Their flavor fades quickly once they are crushed or ground.

Black peppercorns are harvested when the fruit is still green, but nearing ripeness, about the time a single berry shows a blush of yellow. The clusters are threshed to remove them from their tiny stems, and then they are set out on straw mats in the hot sun, where the outer layer shrivels and turns black as it dries. (Sometimes they are dried on their spikes, then threshed.) The peppercorns take many days to become completely dry; they are combed with a wooden rake each day. Once dry, they are packed into sacks and sold to exporters, who trade them on the world market.

Fruit for white pepper is picked about a week after fruit for black pepper. It is a little more ripe—maybe a few berries show that yellow blush, maybe a single berry has turned reddish orange—but not fully red; to wait that long, pepper farmers tell me, is to extend an invitation to the birds who love the ripe berries (birds do not taste heat; some bird seed manufacturers add cayenne to the mix to discourage squirrels, rats, and other mammals who are sensitive to the fire). After the almost-ripe berries are picked, they are tied in big sacks and thrown into the nearest body of water—pond, lake, stream, or river—where they soak for several days. Soaking rots the outer mantle of the peppercorn, the skin that otherwise turns black; what hasn't already come off when the wet peppercorns are emptied from the sack is rubbed off before the peppercorns are spread on straw mats, raked, and dried in the sun, just as black peppercorns are. White peppercorns are traded similarly to black peppercorns, through exporters and traders on the world market.

There are two other pepper products that deserve note. Sometimes the dark skin of black pepper is removed to produce something that resembles white pepper visually, but not in taste. This pepper is called decorticated black pepper (and, occasionally, decorticated white pepper) and it tastes like other black pepper, yet is not as aromatic. Sometimes small quantities of immature pepper spikes are harvested before the peppercorns begin to grow; these spikes are preserved in brine. You occasionally see them in Asian markets.

Citizenship of Pepper

Pepper is classified according to the country in which it is grown. Thus, Malabar pepper is grown all along the Malabar coast of India, in the state of Kerala. Tellicherry, the name of the port that was once the origin of most pepper shipments, indicates a different grade of Indian pepper, larger berries with a greater quantity of volatile oils and piperine and a distinctly floral aroma. Pepper is assumed to be indigenous to this region (although some sources point to the East Indies, currently Indonesia, as the native home of *P. nigrum*) and India was long the leader in quantity, though in 2013 it fell to third place, with Vietnam the top producer and Indonesia second. The top five producing countries, Vietnam, Indonesia, India, Brazil, and China, seem to play a game of musical chairs, reversing their order year by year, or nearly so.

The price of peppercorns fluctuates in ten-year cycles. When there are a lot of producers and plenty of pepper, prices decline, farmers turn to other crops, production drops off, and prices begin to rise. Currently, the price of pepper, which is traded on the world market, is at an historic high and there is a scramble to replant fields that were either abandoned or planted with other crops.

In late winter 2015, the peppercorn market was just coming back after the Tet—Lunar New Year—holidays in Vietnam. The Southeast Asian country has, for two decades, been the world's largest producer of black pepper and so when it pauses, the market pauses.

Why has Vietnam come to dominate world pepper? It is an agricultural society, with a communist government that subsidizes farmers. Thirty years ago, government officials told farmers to grow coffee and the country is now the world's largest producer. Next came the directive to grow cashews; before long, Vietnam became the world's second largest producer.

Faux Peppercorns

Although not true peppercorns, pink peppercorns turned up in the late 1970s on trendy menus and in gourmet markets everywhere, an affectation of the nouvelle cuisine frenzy. But by the early 1980s, reports of adverse reactions eclipsed their chic status; for a time, their sale was prohibited, but the restriction was rescinded through the efforts of French producers, backed by their government, determined to reopen the market for pink peppercorns from the French island, Réunion, in the Indian Ocean, the source of nearly all of the small, brittle berries.

Pink peppercorns come from a tree (*Schinus terebinthifolius*, sometimes called the Brazilian pepper tree) native to South America. A papery outer layer—hot pink, hence the name—shields a small, hard seed with a mildly sweet, mildly aromatic flavor that does not resemble true pepper in the least. Many spice companies include them in four-pepper blends, in which they contribute color but little if any flavor. You see them on a lot of restaurant menus, too, in dishes like salmon in a pink peppercorn crust. I find both their taste and their papery texture distracting, especially when used solo.

Two decades ago, Vietnamese farmers were told to grow peppercorns and now they stand on top of the fragrant pile.

Many major producers have turned to alternate crops because of Vietnam's dominance. Other changes have shaken up the world market, too. In both China and India, a growing middle class has resulted in an exponential increase in domestic consumption of pepper. China has always consumed most of what it produces and now India is more of an importer than an exporter. Indian production has also declined because farmers find it difficult to hire enough workers. Brazil's production dropped off significantly from a high of about 50,000 tons a year but production is increasing once again; most of Brazil's harvest goes to Europe.

Indonesia may or may not be the top producer of white pepper in the world but it is certainly the best known. Much of its pepper is grown on the nearby island of Bangka and is shipped through one of two ports on Sumatra, though it is the Port of Muntok that gets the credit; all Indonesian white pepper is known by the moniker. Black pepper from Indonesia bears the name of Lampong, a port city on the island of Sumatra. Nearly all Indonesian pepper is exported.

In 2015, members of the International Pepper Community included Brazil, India, Indonesia, Malaysia, Sri Lanka, and Vietnam, the trade organization for major producers. The savory vine is grown in many other countries, too, none of which produce enough to trade on the world market.

The United States remains the largest consumer of pepper in the world. In 2015, we imported 50,000 tons of whole black peppercorns, 5,000 tons of whole white peppercorns, and 15,000 pounds of ground white pepper.

"Quality absolutely does not come into play when U.S. traders buy pepper. Cleanliness is an issue but quality is not," he says, confirming what I have heard from other sources, that the United States is not a good market for premium pepper. By cleanliness, Goldsmith means an absence of stems and other plant material that might be mixed in with the peppercorns. We are the only country that sterilizes pepper using ethylene oxide though there is some steam

sterilization of pepper now, too. Irradiation is used, as well, but only for pepper used in the food service industry, not for pepper sold on the retail market.

The Question of Quality

To say that U.S. traders are not concerned with the quality of pepper sounds more alarming than it is. In a sense, pepper is pepper, as salt is salt, its major characteristics of heat and aroma always present with only subtle variations. Quality is more a matter of degree and intensity. Volatile oils contribute to pepper's heady aromas; the alkaloid piperine oleoresin accounts for the heat and flavor. These elements blossom as the longer clusters of berries hang on the vine (to a point, that is; left too long, they begin to fade), so if a farmer is willing to be patient and take the risks involved with later harvest, the pepper will be more peppery, more itself. The aroma and flavor of pepper can be degraded, too, if it is contaminated with bacteria and the toxins they produce.

There is another aspect to the story of peppercorns than the one demonstrated by major spice traders. Smaller companies, some of them mail-order suppliers, some of them selling wholesale to the public, do search for high quality and sell spices that are neither chemically sterilized (the United States is the only country in the world still using ethylene oxide to sterilize spices) nor irradiated. Some manage to procure spices that have been grown without chemical fertilizers and pesticides, no easy task with products from the tropics, where farmers are reluctant to risk losing crops to the pests and viruses that thrive in a humid climate. Yet organic farming is on the rise around the world, including in developing countries and the tropics. Malaysia and Sri Lanka, for example, are devoting considerable resources to the organic farming of both pepper and other spices.

Cat City

"It's easier to find out what's going on in Malaysia," Bill Penzey of Penzeys Ltd. tells me while I'm working on this book, adding that there are interesting developments that will make Malaysian pepper increasingly important in the coming years. I've visited southern India already and understand well the physical and cultural challenges I would encounter if I pursued the pepper trail along the Malabar coast. I welcome the chance for discovery, and I recognize the name Borneo from long-ago Sunday school classes. (*But where is it exactly?* I ask myself.)

"You should stay at the Holiday Inn," he continues. "Room service makes a great steak au poivre with fresh green peppercorns."

Two months later, I arrive in Kuching, the capital of Sarawak, one of two Malaysian states on the island of Borneo (the largest portion of which belongs to Indonesia). In front of the hotel is one of several large cat sculptures in this city whose name means "cat," this one showing a litter of playful kittens surrounding two adult felines. The city celebrates its namesake not only with whimsical statues but also with one of the world's only cat museums, where T-shirts and tea towels proclaim Kuching "Cat City." Other cities in Sarawak pay tribute to signature agricultural products—Serian has a statue of the infamous durian (the "King of Fruits" they say, and it must be seen, tasted, and most of all, smelled, to be believed) in its town square; Sibu, in the heart of pepper farm country, has a statue of a pineapple, also an important crop.

The next morning, I head to the Pepper Marketing Board, a federal agency, founded in 1972, responsible for grading and certifying all pepper exported from Malaysia. It also markets the spice, and processes between 10 and 15 percent of the pepper grown in the country. Modern pepper cultivation was underway in 1875 in Sarawak, then a British colony; by the 1930s, it was producing a significant quantity of the spice. It wasn't until the 1950s, however, that quality and production became consistent.

Tropical air, heavy and saturated, is a fragrant soup of aromas, some appealing, some staggeringly offensive until you become used to them. As I approach the Kuching office of the PMB, the seductive aroma of black pepper mingles with the other smells. The air is thoroughly intoxicating and suddenly, I'm thinking of lunch.

"Black pepper stimulates appetite, making you salivate," Anandan Abdullah, then general manager of the PMB, volunteers, aware of the nature of the commodity he has overseen for more than two decades.

As he shows me through the warehouse where pepper is received, packaged, and shipped, the aroma of black pepper eclipses everything until I am frantically, savagely hungry. *Feed me now!* I think, but I'm quiet, a polite visitor. (Could black pepper oil be used to stimulate the appetites of anorexics and cancer and AIDS patients, I wonder. I've not found an answer.)

Anandan apologizes for spiders I don't see. They remove webs once a week, he says, but let the spiders live; they are very successful at controlling mosquitoes and other insects. Cats, currently sleeping out of sight, keep the facility free of the rats who would eat the jute bags that contain the pepper. Environmental sensitivity appears to underscore everything here.

I taste several types of pepper, crunching into whole peppercorns with abandon, then cringing as their heat sears my palate. Organic black pepper, from a small test farm, is remarkably fragrant, sweet, and fresh tasting. Creamy White Pepper, with its large, evenly sized berries and a uniform pale color that does indeed resemble cream, is exceptional and is sold as one of two value-added peppers for which the PMB is trying to create a

Am I Making You Hungry?

Pepper is an important ingredient in the fragrance industry. Through distillation, volatile oils are extracted from peppercorns (almost always black, occasionally white). It contributes to a perfume's middle notes (those that dissipate within two or three hours; top notes vanish within about twenty minutes; bottom notes linger for five or six hours), providing a warm, earthy element. Polo includes black pepper; Modern Eau de Parfum, released by the Banana Republic in 1998, uses white pepper.

consistent market so that when the price heads downward from its present high, farmers will have a safety net.

In the summer of 1998, world pepper prices are high and farmers are happy; prices have been rising for several years. But pepper prices fluctuate in eight- to ten-year cycles, and they are bound to fall sometime soon. When the price is low, as it was in the early 1990s, fewer farmers are motivated to grow pepper. Some replant with more lucrative crops; some stockpile their pepper hoping for a price increase; some ignore their vines because the compensation is so meager. Gradually, supply decreases, there is not enough pepper to fill world demand, prices begin to rise, and farmers become interested again. They clean out the warehouse and plant new vines (which take a year and half until their first harvest, three years to hit their stride); increased production moves slowly and for a while prices continue to rise. If the PMB can create a market for Malaysian pepper, specifically for Creamy White Pepper and Naturally Clean Black Pepper, farmers will be able to ride out these economic cycles.

The next morning, I begin the long, winding ride to Sibu, 462 kilometers northwest of Kuching, on the Rajang River in the heart of the Sarikei Division (that is, county), where 40 percent of Sarawak's pepper is grown. First, we stop at Serian to observe the process of making Creamy White Pepper.

Creamy White Pepper begins with large berry clusters; the PMB pays a premium for quality fruit. Instead of being tied in sacks and placed in a stream or river, the peppercorns are placed in large plastic barrels (covered), where fresh water from a nearby mountain spring circulates through them for close to two weeks. After they are drained, rubbed clean, and thoroughly dried, they are sorted to remove dark, small, and light (by weight) berries. Most of the pepper is sold to Japan and western Europe.

"The U.S. is not a good market for high quality," Anandan explains. "It is a bulk market, the biggest user of pepper in the world. Most white pepper is used by sausage makers and food packers; they want consistency and are not looking for high quality." (ASTA, the American Spice Trade Association, specifies moisture content, cleanliness, and allowable percentage of light berries; Malaysia packages a Brown Label pepper—to meet these requirements.) Of the more than 70,000 tons of pepper—in all its forms—imported by the United States in 2014, only 140 million tons came from Malaysia in 2014 and none of it was Naturally Clean Black Pepper or Creamy White Pepper. Nearly all of these two peppers are sold to Taiwan, Japan, and Korea. The best way to get it, Marty Goldsmith confirmed, is to find a friend in one of those countries who will send you some.

There are forty thousand hectares of pepper farms in Sarawak, and forty thousand pepper farmers. The average farmer has between two and three hundred vines, which are planted six hundred to the acre. A good vine produces between four and six pounds of peppercorns a year, a great vine as much as ten or twelve pounds; it will remain at peak yield for fifteen to twenty years. As we drive through the Sarikei Division toward Sibu, the hillsides are studded with small farms, some with mature vines in full production, some with tiny new vines covered with dried ferns to shield them from the sun.

Bintangor is a small agricultural town near Sibu. It is the location of the first modern processing plant in Malaysia for Naturally Clean Black Pepper, the other value-added pepper developed by the PMB. Here, fresh peppercorns arrive one day and by the next are ready to be shipped. What normally takes weeks is condensed into twenty-four hours.

Sarikei pepper farmers deliver their pepper in the early morning, before the heat of the day. The Pepper Marketing Board buys the pepper, stalks and all, before it is threshed, with the requirement that it be delivered to the facility within twenty-four hours of picking.

Shortly after its arrival at the facility, the pepper enters the first stage of a semiautomated process, a double thresher that removes virtually all plant material before sending the berries through a cold-water bath to wash away dirt and dust. The washed berries go into a one-minute hot-water rinse to kill microbes and are then carried up a conveyor belt into an enormous flat-bed

stainless-steel tank, where huge fans blow hot air over them for twenty to twenty-four hours, during which time their skins wrinkle and turn black.

When I visit the center late on a hot June morning, the fans have recently been shut off. I stand on my tiptoes and poke my head into the dryer, whose doors are open to cool the pepper, breathing in the luscious dark aromas. *When's lunch?* I think instantly.

Once cool, the dried berries enter another conveyor system, passing through a blower that separates the dust from the berries on their way to being packed in twenty-kilo plastic sacks that will be slipped inside three-ply paper bags and labeled. The pepper will be sent to the Kuching facility to be shipped. (It is a confirmation of the increasing popularity of pepper that the dust, once discarded, is now sold; it makes an excellent mulch and fertilizer in your garden, I am told. I wish I could get some back home.)

Wong Hoon Sing, grading inspector for the PMB, tells me this is the best pepper in the world, an assessment I don't feel qualified, at the moment, to confirm. The newly dried pepper looks beautiful—clean, without stems or dust, evenly sized—and smells clean, earthy, rich, and erotic all at the same time. I crack a peppercorn between my teeth and my mouth is flooded with heat and flavor. I have never had better pepper, but then, I have never tasted it this fresh, nor seen it in this quantity before. Is my judgment eclipsed by my exotic surroundings?

At this time in history we look toward old, traditional methods of food production as superior to our modern, industrialized techniques. We search out the hand-crafted, the sun-dried, the pure and simple. We want heirloom vegetables, artisan cheeses, condiment olives from small farms, range-fed organic beef from steers who get daily massages. Yet here, on the edge of the rain forest on the island of Borneo, I am enchanted by pepper made in a completely new, technologically sophisticated method. Most pepper is sun-dried by the farmers who grow it, raked with the same wooden rakes they've used for decades, for centuries. What is the story?

The Queen of Spices

If pepper is the King of Spices, Malaysia's Naturally Clean Black Pepper is its queen. At the time I first tasted NCBP, I was enchanted but inexperienced. I had never tasted better pepper but I thought, perhaps, I might. I was wrong. Naturally Clean Black Pepper is, without question, the finest pepper in the world, with subtle sweet qualities and engaging floral aromas that other peppers lack. You can, should you be so inclined, say it has a feminine nature; should you prefer musical analogies to gender comparisons, think of it this way: NCBP entwines the bass notes of all pepper everywhere with a singular and delicate melody, unique to this pepper.

When I returned to Kuching a few months after my first visit to speak at the International Pepper Festival, I brought about fifteen pounds of NCBP home with me. It lasted a few years and then I began buying it, pounds at a time, from Penzeys, until they ran out. I have a few ounces left, enough to fill a pint jar, and I keep it near me so that I can unscrew the lid, lean close, and breathe in its seductive aromas, punctuated by those bright floral notes, the signature quality that other peppers lose, degraded by lengthy drying, by toxins, by chemical sterilization. I try every Sarawak black pepper I can find, I search the Web, I send emails and make phone calls, and I have not yet found a domestic source for this precious peppercorn. Still, I try, every time I discover a new spice vendor. I suspect I may need to make a third trip to Kuching.

In fact, the alliance between the PMB and the local farmers seems to be an extremely successful union between the old world and the new. The PMB has a vested interest in the success of the farmers, and the agency seems to make a farmer's life easier—they are paid a premium for good farming practices that result in quality fruit, and they can deliver their crop shortly after harvest.

There is another element to consider. All pepper is tested for bacteria, including salmonella and *e. coli*, its two most common contaminants and the main reason pepper is sterilized and irradiated. Pepper is particularly vulnerable to such contamination because it sits for several weeks on those straw mats on the ground in front of the family home, where chickens run free and birds fly overhead. If the wind comes up, as it always does at some point, dirt, chicken droppings, and a multitude of other pesky little things sweep right over the drying pepper, some inevitably dropping into it. As resistant strains of bacteria develop, there could be a serious threat of dangerous contamination.

Certainly, bacteria are killed during sterilization, but by then toxins released by bacteria have already degraded the flavor of the pepper. Chemical sterilization has the added liability of introducing harmful chemicals into the environment. By processing the pepper before bacteria contaminate it, the potential for flavor damage is eliminated; the water bath kills bacteria and the pepper needs no further sterilization.

For pepper that does not meet the requirements for Naturally Clean Black Pepper, the FAQ (Fair to Average Quality, the world standard) pepper, which the PMB also processes, there is a new steam sterilization plant at the Kuching facility. The single drawback is that steam sterilization is expensive; the chemicals used in sterilization are inexpensive (but only in the short run, and only because the companies that produce them want them used). Other countries have similar plants, but the total amount of pepper (and other spices) processed this way remains small.

Rice Wine and a Crescent Moon

The population of Sarawak is about 51 percent Iban, a colorful native tribe with centuries of history on the island. Their traditional festival costumes, dances, and music are among the most beautiful and compelling I have seen anywhere. Today, many Iban are pepper farmers who live in longhouses, traditional structures (in some cases built with modern materials; in some cases with thatch, bamboo, and wood from the rain forest) with a dozen or more families sharing a single extended roof.

Before we begin the long drive back to Kuching, we visit Kotak ("box") longhouse, not far from the PMB's Bintangor processing plant. A narrow porch leading to the main house is crowded with black and green peppercorns drying in the sun; more are on the ground outside the house. Inside, the common area, which stretches the length of the twenty-two-door structure (a longhouse is measured by its number of doors, each of which leads to a single-family apartment; think of it as a horizontal apartment house, with an enormous lobby shared by all tenants), is deserted.

The chief greets us and slowly a few residents join him, curious about their redheaded visitor. Older men smile and stare; young men laugh. Women approach more slowly, bringing sweet snacks but always hanging back, never joining the circle where we sit with the chief. A young man pours me a glass of rice wine, a traditional beverage made by each family; if I were here on a feast day, it would be rude not to accept (and drink) a glass from each family. The wine is smooth and refreshing, but deceptively potent.

One of the snacks is a large, airy thing of grain and sugar resembling a crescent moon made of spun gold. When I hold it up as if in the sky, everyone laughs and it feels as if we have bridged a cultural gap, the inevitable uncomfortableness between strangers. We've all watched a crescent moon rise in the night sky, we all recognize it for itself. They laugh again when I accept a second glass of wine. They ask, through Wong, if I would like to spend a night. An overnight stay at a longhouse is a popular tourist activity; some build special guest houses to encourage the additional source of income. My spontaneous "yes" draws another shared chuckle. I am so obviously enchanted by my hosts and I am sorry when it is time to go.

Each family here has a pepper farm. Before leaving, we walk to the nearest ones, just across a road that is both muddy from daily rains and dusty from baking in the sun. The noon sun is so hot I feel faint, but soon I am partially shielded from its full intensity by towering pepper vines, currently under harvest. A sack half full with just-picked pepper clusters lies between two rows. I pluck off a berry and chew it; it is spicy hot, aromatic, and pleasantly crunchy. It tastes just like black pepper, only greener, brighter, more urgent. Here, too, the air is a thick stew of peppery aromas. Unlike, say, vanilla, pepper does not need processing to develop its qualities; they are present from the start.

Two farms border a pond where white pepper is soaked, protected from the muddy bottom by wooden platforms. When I listen carefully, I can hear water from a nearby stream trickling into the pond.

Some Kotak farmers sell their pepper to the PMB, others dry it themselves and sell to exporters. Transportation is not a problem here; this longhouse is close to Bintangor and Sibu, but not all pepper farms are so conveniently situated. Hundreds are several hours away and farmers must travel first by boat and then by car, toting their harvest, to deliver their pepper. The PMB employs teams of traders who visit remote farms, buy the pepper, and bring it back to be processed. If a farmer can sell his pepper easily, he is more likely to keep growing it.

Interestingly, pepper doesn't play much of a role in the local cuisine, Iban or otherwise. Whenever I comment on what seems to me an unusual omission, I receive the same response: "Our food is so fresh and so good, we don't need pepper." The Malaysians I spoke with think of pepper as something to add flavor when flavor is lacking (and as a source of income, of course), and to some degree, as a preservative. When they use pepper, it is usually white, ground into a fine powder. There are exceptions, of course. A multicultural country, Malaysia has substantial Indian and Chinese populations, as well as Nyonyas, as those of native Malay and Chinese ancestry are called. Indian cuisine makes substantial use of black pepper, as does Chinese, though to a lesser degree. Nyonya cuisine includes a number of traditional dishes that rely on black pepper for traditional flavor.

In 1991, the PMB launched the annual Pepper Festival, held in October in Kuching and supported by the city's hotels and hotel chefs, who develop special pepper recipes. The board also is attempting to increase the use of pepper through the development of a number of commercial

An Aromatic Aphrodisiac

In the master bath of the Marilyn Monroe Suite in the Legend Hotel (where Windows 95 was launched, by the way) in Kuala Lumpur, Malaysia, there is a small ceramic aromatherapy bowl, set over a votive candle. When the lid is removed, the air-conditioned room is filled with the mingled aromas of tangerine, black pepper, and sandalwood. Near the black-tiled sunken tub, there's a beautiful bar of soap scented with patchouli, black pepper, and ylang ylang; it's called "Sexy." (Other blends are known as "Uplifting," "Refreshing," "Relaxing"; which would you choose?)

It is coincidence that I am here, here being an aromatherapy seminar conducted by two certified practitioners from Culpeper Ltd., a small company based in London that produces aromatherapy oils, massage oils, soaps, and a few other luxurious products for skin and spirit. Black pepper, they emphasize, is very strong and is used only in small amounts in two blended massage oils, one to relieve sore muscles, and one with the same name as that soap, Sexy. Black pepper, like many substances that stimulate the senses, is believed to be an aphrodisiac. I return to California with both oils as well as pure black pepper oil, which I burn in my aromatherapy pot, an attempt at inspiration, an urgent call to my Muse. It works, apparently; you're reading the results.

products, including candy made with black pepper oil. During my stay at the Holiday Inn, I found a piece of pepper candy on my pillow each evening, as sweet and spicy as a goodnight kiss, or almost.

A Brief History of the World

Exploring the history of pepper is akin to taking a refresher course in the lessons of elementary school. The names and stories are so familiar. There's a confused Christopher Columbus stumbling onto the West Indies, there's Magellan and Marco Polo, trade routes, and overland spice trails. There's Portugal's famous navigator, Vasco da Gama ("For Christ and spices!" he is said to have shouted upon landing in India), and the Dutch East India Company. And there are all those annoying pirates too—pirate-infested waters, one writer puts it, as if they were sharks or rats—tripping them up at every unexpected jut of land, every swell of the sea. How did they ever make it?

And weren't we told they did it all for *spices,* especially for black pepper? That's taking the short view, don't you think? There are no spices on the moon, but we couldn't resist making that effort, could we? If it hadn't been spices, it would have been something else. Both the exploration and the spices were inevitable. It was merely a matter of time.

Sanskrit texts refer to pepper as the first spice, though it took a while to make it to Europe. Homer doesn't mention it, but by the fourth century B.C., Theophrastus, a Greek philosopher, described both *Piper nigrum* and *P. longum,* the long pepper that was initially more popular in Europe than black pepper was. Pliny the Elder, writing in Rome in the first century A.D., reports that long pepper cost four times as much as black pepper did. When Rome imposed a duty, it was on white pepper and long pepper; black pepper was spared, apparently not fashionable enough to warrant the effort.

An elaborate spice market was built in Rome, and the road that led to it was Via Piperatica, Pepper Road. Spices, so expensive and so sought after, were heavily guarded. From Rome, all three peppers (white, black, and long, there was no green then) spread throughout western Europe.

Peppercorn Rent

In England, rent was paid with a pound of pepper, a payment occasionally made today as a symbolic recognition of property rights; Prince Charles receives peppercorn rent from Wales. British contracts still contain a "peppercorn" clause, and the gesture is considered an acknowledgment of ownership.

Pepper Licks

"Go where the pepper grows," a German irritated by a neighbor might say: *Get away from me, and go far.* For annoyances closer to home, pepper has long been believed an effective carminative (though no one says how much you must eat). It can still both vertigo and nausea, and will improve digestion in general. If you need relief from laryngitis, make a paste of finely ground white pepper and butter. Take a lick every now and then until your voice is restored.

—Carolyn Heal and Michael Allsop
Cooking with Spices

It is not clear exactly when black pepper overtook long pepper in popularity, but peppercorns were used abundantly in Rome to season everything from sauces to desserts. By the time the Visigoths conquered the city in 408 A.D., black pepper was so dear that they demanded three thousand pounds of it as part of their tribute (the rest was in gold and silver). You could say it served the Romans right, surrendering all that pepper. After all, they're the ones who, upon destroying Carthage (on the north coast of Africa about ten miles east of the modern Tunis) in the second century, covered the city in salt, rendering it uninhabitable, as they left. You reap what you sow.

After Rome fell, Europeans frittered away their time for centuries and pepper became scarce once again. And, like so many rare things, its value grew and grew and grew, surpassing any reasonable price you might normally pay for a seasoning. Pepper became so valuable, so desirable, that you could use it as money to negotiate any debt. It was considered to be even more stable than gold, because the gold on coins could be scraped off by the unscrupulous. Just as balsamic vinegar would be a few centuries later, peppercorns were passed on through dowries and the estates of the wealthy.

In the medieval world, wealth often was expressed not by the amount of gold in one's vaults but by the amount of pepper in one's pantry. It is said that both rents and salaries were paid with peppercorns and that a slave could buy freedom with a pound of them, though you might wonder where the slave would get the peppercorns in the first place. In France, pepper was a customary bribe to judges, and "pepper bags" a derogatory moniker in Saxony for noblemen who married commoners for their money.

There is more to the story of pepper at this time than high prices; great mystique surrounds it, and plenty of misunderstanding too. My favorite story is one told by Waverley Root in *Food*. "Pepper ripens in the heat of the sun," a fourteenth-century friar is reported to have explained, "and serpents defend the woods where it grows; to pick it, the trees are set on fire [to drive the serpents away] and the pepper becomes black." It was centuries later that the American nutrition fanatic Sylvester Graham claimed that excessive use of pepper would cause insanity.

Eventually the Crusades would get Europeans up and moving, but when it came to spices, the fervor was fueled by economic competition and religion more than the pursuit of flavor.

During this period, Venice, having secured its position and fortune with salt harvested from nearby marshlands, controlled trade within Europe. The Venetians sailed to Alexandria in ships filled with soldiers and returned loaded with spices, including pepper. The missing link was how the spices got to the Holy Land in the first place. The Arab spice monopoly had been in place at least since 950 B.C. and Arabs had been successful at keeping their overland trails and sea routes secret, but European explorers became determined to make their own way to the cache of tasty barks, berries, and seeds in the tropics.

Genoa, also a thriving port but a greater distance from Alexandria than was Venice, posed the only competition until the Portuguese reached India by sea in 1498. Then all bets were off; the Europeans were no longer dependent on the Arabs for peppercorns or other spices. The Portuguese remained dominant for a time, reaching the East Indies, including the Spice Islands, before others. Then the Dutch wrested control from

Portugal, and eventually the British usurped their power. Spain competed briefly, but found the gold in North America more alluring than spices; they sold their rights along the spice trail to Portugal.

By the late 1700s, the United States was involved. Elihu Yale was the first colonist to get into the pepper trade, and the fortune he made went to establish Yale University. In 1778, when Jonathan Carnes of Salem, Massachusetts, acquired direct trading partners in Southeast Asia, including some in Sumatra, then the source of the world's best pepper, Salem became the Venice of the New World. America's role in the spice trade grew, and for a time taxes raised from imported pepper paid 5 percent of all U.S. government expenses. In 1805, we reexported 7.5 million pounds of black pepper. In spite of the difficult journey—the route from Salem to Sumatra and back was twenty-four thousand miles by sea—the industry thrived. In 1873, U.S. ships made a thousand round trips.

By this time, the transcontinental railroad had been completed. It became customary in the dining cars to offer pepper at the table, in pepper grinders, so that diners could see that it was indeed black pepper, and not bits of coal, seasoning their food. There was reason to be suspicious of unidentified black specks. The popularity of ground pepper in Europe had inevitably given rise to unscrupulous suppliers. In England in the 1700s and 1800s, ground black pepper was often adulterated with a variety of cheap materials, from ground olive pits and date pits to crushed dried leaves and the ground hulls of black mustard. White pepper was mixed with ground rice. By 1900, James Trager tells us in *The Food Chronology*, most ground pepper sold in London contained very little pepper. The adulteration was so common that actual pepper might be returned to the market as too strong. So much for the pursuit of flavor.

PART III
A Salt & Pepper Cookbook

Seasoning to Taste

The amount of salt and pepper in most recipes, except for baking, has been left, since the earliest cookbooks, to individual taste. The casual simplicity of the instruction "to taste" is appealing and for the most part good advice. The balance of salt and pepper in most savory dishes should be adjusted just before serving. A flat sauce will perk right up with a little salt; a dull soup will blossom with flavor when the right quantities of salt and pepper are added. Everything is improved when it is seasoned properly and the only way to accomplish this is by tasting.

For foods to reach peak flavor, you must salt as you cook, as well—onions as they sauté, for example. Think of cooking as a construction project; each step needs to be completed before you go on to the next. Add salt and pepper in stages, and then taste the finished dish, which will need at most a bit of adjustment. This technique is easiest to master when you salt with your fingers; you will quickly develop an intuitive feel for the correct amount. A sturdy pepper mill is best for adding pepper. If you build flavor in this way, your cooking will improve.

Vegetables should be blanched in salted water, and the water for boiling them should be salted, too. Pasta cooked in unsalted water will never be as flavorful as that cooked in salted water, no matter how tasty the sauce or how much salt you add at the table.

Recently I've heard a few cookbook authors, food writers, and chefs insist that salt should be added only at the end of a recipe and that this is the best way to reduce the amount of salt used. This is the opposite of how the best home cooks and best chefs cook and it is bad advice. Ignore it. Suspicious but curious? Do a taste test: Cook two batches of pasta, one in generously salted water, one in unsalted water, or prepare two versions of a dish, one salted during cooking, one salted only at the end. Let your palate judge.

In the United States, we usually think of white pepper as pepper without color; we use it in pale sauces to avoid black specks. White pepper is almost always more expensive than black and so we intuitively use less of it. In Europe and much of Asia, the reverse is true. White pepper is less expensive, and it is considered the all-purpose pepper for cooking. In Europe, pricier black pepper is added at the table, as a condiment.

There is an element of wisdom in this tradition, though I don't believe the differences are so significant that they warrant using the more expensive pepper. The crux of the difference is that there are more volatile oils in black pepper, and those oils begin to dissipate once the pepper is cracked and continue to do so as the pepper is exposed to heat. If black pepper is expensive, it makes sense to conserve it rather than let an important element of its character vaporize.

Good white pepper has a particular taste that results from the lengthy soaking that removes its outer mantle. It's a deep, almost caramelized flavor that some describe as winy. Because of this almost-sweet character, white pepper goes well with ginger, cardamom, and other aromatic spices. But so does black pepper, whose full range of aromas and flavors makes an essential contribution to most dishes.

A word of caution is in order. Although salt improves the flavor of almost everything, neither it nor its pungent companion pepper can disguise poor ingredients. In the late 1990s, many American chefs began to return to an elemental style of cooking. Paul Bertolli, former chef of Oliveto in Oakland, California, praises the simple goodness of grilled meats seasoned only with salt and pepper. In an article in the *Wine Spectator*, the writer heaps praise upon a chef daring enough to season salmon with nothing more than sea salt. Please, I want to add, let's keep things in perspective. If you begin with a wonderful piece of meat or fish, a silky heirloom tomato, or an eggplant lovingly nurtured every step of the way, plucked at precisely the right moment, and served instantly, the simplest preparation in the world will produce sensational results, especially if they are seasoned with a deft hand. Indeed, they will need nothing more than a little good salt, a bit of fresh pepper. (Witness the phenomenal and enduring success of Chez Panisse, a restaurant in which cooking is based upon this principle: ingredients first and ingredients last.)

If, however, you are using poor-quality ingredients, you cannot expect miracles just because you change the type of salt or pepper you are using. My best advice is this: If you have banned salt from your kitchen, bring it back (and hurry up!) and make sure it's a good one, not table salt. If your pepper comes already ground in a big tin, toss it and start over with whole peppercorns. But pay attention to other ingredients, too. Shop for seasonal produce at farmers markets and forget about such things as tomatoes, peaches, and apricots in January. Get to know your baker, fishmonger, and butcher; find out who sells the best cheeses and start talking to them. If you can't know the farmers and ranches who produce your foods, at least know where those foods were grown and raised. (Find a source for good inexpensive wine and craft beer, too.) It is remarkably easy to cook well if you begin with a pantry and kitchen full of good things. Without good ingredients to begin with, you're fighting a losing battle that neither salt nor pepper can help you win. That said, if you're stuck somewhere without access to good ingredients, skillful application of salt and pepper *will* help.

As you would expect with something as recently liberated as salt, there has been no end to the silliness surrounding its use. (But, hey, better exuberance than persecution.) You can now find both articles and recipes that warn that salt, like pepper, must be freshly ground for optimum flavor. This is ridiculous. *Salt is a rock.* There is no organic material to deteriorate, no volatile oils to dissipate. Very dry salt might absorb a little moisture in a

humid atmosphere; moist sea salt might lose moisture in a dry climate. But its flavor does not change. There is no taste benefit to grinding salt immediately before using it.

The Shaker Dilemma

When you switch from granulated table salt to flake salt, unrefined solar-dried sea salt, or other specialty salts, your shakers become useless. How do you make the salt easy to use, both at the stove and at the table? On my kitchen counter, I keep a rectangular wooden box with a hinged lid; I use one made of maple that holds about eight ounces of kosher salt. Next to the stove, I keep a salt pig big enough for me to reach in with my hand. At the table, salt cellars with tiny silver spoons are perfect at a formal dinner party, but for everyday use, something casual is better; those tiny spoons are a bit effete and impractical. Small ramekins work well, as do small, shallow condiment dishes.

In response to the increasing popularity of coarse salts, several manufacturers have added sturdy salt grinders to their lines of pepper mills. This newer generation of salt mills is an improvement over earlier models, which ground the salt to a fine powder; the new ones have adjustable grinding mechanisms and easily produce coarse flakes. Although I appreciate the

Please Pass the Stethoscope

During the filming of the original *Star Trek* television show, a prop stylist was sent out to buy salt and pepper shakers for the Starship Enterprise. He returned with the most bizarre ones he could find. "They won't do," he was told. They were too strange, no one would realize what they were. Someone went to the studio's cafeteria for standard glass shakers, the one with the little metal lids. The space-age salt and pepper shakers didn't go to waste, though; they became Dr. McCoy's medical instruments, diagnostic tools that he waved over his patients' bodies to diagnose their ailments.

increased efficiency, I prefer to use my fingers to add salt both during cooking and at the table. I usually keep a salt grinder as well as salt cellars on my table so my guests have a choice.

Shake It, Don't Break It

A tiny dachshund that separates in the middle; an alligator in sunglasses; a kangaroo and its joey, both in tuxedos; a gospel quartet; a rising sun; a full moon; a crescent moon and star; cows in a convertible Cadillac; a jukebox; a gas pump; giant ants; a raven atop a skull; copulating bears; kissing kittens; a burger and fries; Elvis and his guitar; Betty Boop in red shoes; Mount Rushmore; Bakelite cherries with sterling silver leaves; a bride and groom, who, when viewed from the back, become an angry couple, her bouquet transformed into a rolling pin. What do these images have in common? All are salt and pepper shakers, sitting on collectors' shelves, crammed into glass cabinets with thousands of other pairs, transported with extreme care to annual conventions all over the country. They are photographed for websites, and books are devoted to them. They are rarely filled with salt and pepper.

There is a substantial subculture of salt and pepper shaker collectors: specialists in Victorian shakers, early American shakers, shakers from the 1930s, or 1940s, or 1950s. Salt and pepper shakers are not only the domain of retro kitsch. In 1997, The Clay Studio in Philadelphia held a show of fine-art salt and pepper sculptures. A company in Oregon shapes beautiful, delicate shakers of blown glass, some of them attached to elaborate glass pedestals wound around with glass serpents. There's something about these two-piece evocations of culture that is endlessly intriguing.

Freshly Ground Pepper

In most cases, peppercorns must be cracked, crushed, or ground before being added to other foods. There are many ways to accomplish this, the easiest of which is to purchase ground pepper, though I don't recommend it. Once crushed, peppercorns lose their aromatic quality fairly quickly; for complete flavor, you should use whole peppercorns. (I do keep a small quantity of commercially ground pepper around for Cajun dishes; it adds essential character.)

Cracking is easiest in a suribachi, a molcajete, or with a large mortar and pestle. For best results, place no more than two teaspoons in the container and then press down firmly and directly onto some of the peppercorns until you feel them crack. Continue until they are all evenly cracked. You can also crack pepper by placing it between two sheets of wax paper, cooking parchment, or the folds of a tightly woven tea towel, and then rolling over it with a heavy rolling pin or pounding it with a meat tenderizing mallet until the peppercorns are just cracked. In a pinch, you can use a hammer.

In general, when grinding by hand it is best to avoid small containers because the peppercorns will simply jump out. There is one exception. Mason Cash, an English company that manufactures a variety of ceramic products for the kitchen, makes a small, narrow mortar with an almost cylindrical pestle. Because the sides of the mortar are vertical, you can grind small quantities of peppercorns and other spices without their leaping onto the counter and floor.

Crushed pepper indicates a coarse texture, smaller than cracked but not as fine as ground. Commercial pepper mills do an excellent job, but you can also use a suribachi or other hand grinder. Just continue to grind until the peppercorns are the size you prefer. For ground pepper, fairly fine but not quite a powder, most commercial mills do a good job in not too much time, but again, you can easily do it by hand. For very finely ground pepper, grind the peppercorns by hand or use an electric spice grinder (or converted coffee grinder, either electric or hand-cranked, on the lowest setting).

The Best Pepper Mills

I've tried out more than a dozen pepper mills and always come back to my trusty wooden Peugeot. It's a tad difficult to fill and it doesn't hold

as much as I'd like, but the grind is easy to adjust and it has produced perfect cracked pepper for more than two decades. I use it for my precious Naturally Clean Black Pepper from Sarawak. I use a brass Atlas grinder for white peppercorns; I like its appearance and its hard metallic feel in my hand. It produces a consistent medium-coarse grind, but it's not easy to adjust (you need a screwdriver, often an impossible hurdle), so it has limited possibilities for all-purpose use. I keep two others, similar to the Peugeot, next to my stove, for a total of four. One contains mixed peppercorns and one is filled with what I call my primary pepper, the one I use most of the time. Exactly what it is varies depending on what I have on hand.

Those ridiculously long pepper mills waved by many an overeager waiter (Wait! Let me taste my food!) have a single function, to reach across the table to the customer in the far corner. In selecting a mill for home use, longer or bigger is not necessarily better (though you do want a large enough capacity that you're not refilling it every other day). Many articles rating pepper mills have appeared over the years and several are worthy of note. *Cook's Illustrated* (March/April 1996) and *Good Housekeeping* (May 1997) both rate the Unicorn Magnum Plus the best mill. It grinds quickly (five seconds to grind half a teaspoon according to *Good Housekeeping*, ten seconds according to *Cook's Illustrated*), produces a wide range of grinds, holds one and a half cups of peppercorns, and is very easy to fill.

A ten-inch maple wood mill sold by Chef Specialties placed second in both evaluations. *Good Housekeeping* also recommended the Zassenhaus grinder, which *Cook's Illustrated* did not test. Olde Thompson Manor pepper mill was recommended by both publications. If you stick to any of these brands, you'll have good results grinding pepper. I don't recommend novelty grinders that offer high-concept designs and technical gymnastics. None improves on the conventional design of a good grinding mechanism made of stainless steel and a sturdy hand crank or bulb.

Several reviews in 2013 show Unicorn Magnum Plus remains the top-rated pepper mill, with Cole & Mason Derwent Precision Pepper

Mill named as a reliable alternative. Among inexpensive pepper mills, Oxo Good Grips Pepper Grinder gets high ratings.

If you want to go high end, you have plenty of options, including a beautiful tall mill called "Aphrodite" crafted in African olive wood, signed by the artist and sold at Pepper-Passion, an online retail store, for two hundred twenty dollars.

A great source for good sturdy mills at bargain prices are garage sales. I've found several, which I've purchased for friends who never seem to get around to buying one of their own.

Repetitive Pepper Syndrome

When I was writing the first edition of this book, I spotted electric pepper mills at the San Francisco Fancy Food Show in 1997, when Chef Specialties, Olde Thompson, and Peugeot were introducing battery-operated pepper grinders. The mills all operate on the same principle—there is a push switch on the top, easily pressed by the thumb, which turns the grinding mechanism and at the same time illuminates the cascade of ground pepper and the food to be seasoned. They seemed so silly, so indulgent and decadent that I started to giggle, but then I thought of my late mother-in-law's painful arthritis, a colleague's carpal tunnel syndrome, a culinary student whose movement is limited to a few weak hand motions. All three have sophisticated palates, but are limited in what they can actually do in the kitchen. The electric pepper mill—which seems so frivolous that I can't imagine them buying it for themselves—offers an easy and tasty solution to one of a myriad problems. It's a thoughtful gift for anyone with limited hand or arm movement.

Today, there are many electric pepper mills on the market with prices that range from about twenty five dollars to well over a hundred, plus the cost of batteries. It is best to choose an electric mill based on an individual's motion restrictions.

There are now electric salt mills, too.

A Pound of Pepper

Recipes often call for five peppercorns, or ten, or thirty. It's helpful to know how the numbers may be translated into standard level measurements. The ranges account for the smaller peppercorns, Malabar, Muntok White, and Sarawak, and the larger, Tellicherry. There are about twenty-five hundred medium-sized berries in a cup of whole peppercorns. A pound of peppercorns contains between nine thousand and fifteen thousand berries.

⅛ teaspoon 5 to 7 peppercorns
¼ teaspoon 13 to 18 peppercorns
½ teaspoon (1 gram) 22 to 37 peppercorns
1 teaspoon (2 grams) 34 to 60 peppercorns
1 tablespoon (6 grams) 106 to 189 peppercorns

Appetizers, Snacks & Nibbles

Think of an appetizer as an invitation, a proffer, an enticement to make us ready—that is, hungry—for the meal to come. Fine restaurants often serve complimentary *amuses gueules*—the Muses of the meal, as it were, come to bless our pleasure—tiny single bites of savory morsels that tickle and delight our palates. Salt may be the most important element of an appetizer; like a kiss, it enchants the palate and leaves us longing for more. Pepper plies a randy trade as well; even its aroma makes us salivate, which in turn increases our hunger. Appetizers needn't be complex, nor bountiful. Slices of a good baguette, brushed with olive oil and toasted, make perfect little canvases for all manner of condiments, from a smear of chutney or tapenade to a slather of salted egg sauce. Gravlax, too, make an excellent way to start a meal. The most important thing to remember is that you are waking up appetites, not sating them.

Popcorn

French Breakfast Radishes with Butter & Flake Salt

Fried Padrones & Shishitos with Lemon & Burrata

Edamame

Deviled Eggs with Smoked Salt & Green Peppercorns

Hard-Cooked Farm Eggs with Smoked Salt & Butter

Bruschetta with Sautéed Greens

Pineapple with Black Pepper & Hawaiian Alaea Salt

Onandaga Salt Potatoes

Grilled Figs with Prosciutto & Black Pepper

Grilled BLT Kabobs

Taramosalata

Roasted Marrow Bones with Artichoke & Green Olive Tapenade

Red Miso Soup with Garlic, Ginger & Noodles

Lemon Pepper Chicken Drumettes

Brandade de Morue

Olives & Almonds

"When I ate my first olive, I knew instinctively I was tasting the heart of the world" (a comment made to Mort Rosenblum and recounted in his book *Olives*). If an olive is the world's heart, its life-giving blood is salt. Salt effectively leaches out an olive's bitter flavors and it's only after this leaching that we can enjoy them. Salt also returns flavor to those olives stripped of every nuance and distinction by lye, which leaches out bitterness more quickly than salt does (twenty-four hours rather than weeks or months) but takes everything else along with it, leaving the olive tasteless. Lye-cured ripe olives, baptized in a salty brine, become the ubiquitous California-style black olives. A bowl of olives—a mixture of green and black, dry-cured and brine-cured, seasoned with lemons, oranges, chiles, herbs, or nothing at all—is an effortless appetizer that is always pleasing and always appropriate, no matter what follows.

Almonds, especially Marcona almonds from Spain, also make an excellent appetizer, served on their own, with olives, with a few shards of Manchego, a bit of jamon serrano and maybe some membrillo. They are rich in natural oils and salt makes their flavor soar.

Popcorn

Serves 4 to 6

Popcorn is one of the most simple and satisfying nibbles there is. You find it around the world or nearly so, often with nothing more than butter and salt but sometimes, as I learned in India, with spices or condiments popular in that particular region. On a sultry August night in Poona, India, I went to see Franco Zeffirelli's Romeo & Juliet, *screened in a large tent without sides. An adjacent, smaller tent offered refreshments: popcorn and hot chai, served in heavy porcelain cups. My mouth was on fire with my first bite of popcorn, seasoned heavily with salt and ground cayenne. The hot chai, also spicy, did not help cool my palate. Soon, I learned to love the combination of flavors and now almost always put something spicy on my popcorn. A theater in the town where I live has both real butter and an array of condiments—several types of hot sauce, grated cheese, black pepper—for their popcorn.*

Have you ever wondered why corn pops? Certain varieties contain water within their kernels and when heated, this water turns to steam, expands, and must escape. Voilà! Popcorn! You should leave the pot's lid ajar so that this steam escapes; if the popped kernels reabsorb it, they will become tough.

2 tablespoons olive oil
½ to ⅔ cup popcorn kernels
1 cube (4 ounces) butter, or less to taste
Tapatio, Cholula, Sriracha, Crystal, or Tabasco sauce
Kosher salt, lightly crushed smoked salt, or a favorite flavored salt

Set a heavy, wide, deep (4 inches) pan over medium-high heat, add the oil and the popcorn, and watch until the first kernel pops. Quickly cover the pot with its lid, leaving a bit of a gap so that it does not fit tightly, and shake the pan gently every few seconds. As the sound of popping increases, shake gently every few seconds. As the sound begins to taper off, shake less frequently and remove the pot from the heat when there has been no popping for 15 to 20 seconds. Let rest another few minutes, cover off center, during which time any popped kernels should pop.

Meanwhile, put the butter into a small saucepan over medium heat, add a few shakes of hot sauce to it. When the butter is fully melted, remove from the heat.

Transfer the popcorn to a serving bowl, drizzle the butter over it, season generously with salt, turning the popcorn several times, and enjoy.

Variations:

- If you have a favorite seasoned salt, either homemade or commercial—caraway salt, chipotle salt, fennel salt, etc.—use it in place of kosher salt.

- Add finely grated cheese—Dry Jack, Parmigiano-Reggiano, Pepato—to the popcorn after adding salt. Use or omit the hot sauce, based on your preference.

- If you love lavender, omit the hot sauce, combine a tablespoon of ground culinary lavender, a tablespoon of granulated sugar or honey, 2 teaspoons of black pepper, and 2 teaspoons of kosher salt, add to the butter and toss with the popcorn.

French Breakfast Radishes with Butter & Flake Salt

Serves 4 to 6

Here's one of the easiest and most refreshing appetizers you'll ever make. Nothing could be simpler. For a more elegant presentation, consult the Variation at the end of this recipe. If you can't get French Breakfast radishes, use the smallest radishes available and make sure they are not pithy inside.

18 small French breakfast radishes, washed, all but 2 small leaves removed
Small scoop (about 1 tablespoon) unsalted butter, preferably organic
Maldon Salt Flakes or Murray River Flake Salt

If the radish roots seem tough, snip them off. Dry the radishes on a tea towel and transfer them to a serving platter. Add the butter alongside and put a little pile of salt next to it. Serve immediately, with a small knife. Guests either dip the radishes into the butter or use a knife to spread some on the radish before dipping it lightly in the salt.

Variation:

Butter-filled Radishes with Sel Gris: For a more formal presentation, use a sharp knife to make a lengthwise ⅛-inch-deep cut in the side of the radish, holding the knife at a slight inward angle. Make a second cut about ⅛ of an inch from the first, slicing a small wedge out of the radish. Use a butter knife to fill the slit with butter and use the knife to smooth it evenly. Set each filled radish on the bed of radish leaves or parsley. Cover loosely with plastic wrap and chill for 30 minutes, or until the butter is set. Sprinkle a little crushed sel gris on the butter in each radish and serve immediately.

Fried Padrones & Shishitos with Lemon & Burrata

Serves 4 to 6

Like so many other foods—fresh soy beans, farm eggs, and smoked salt, for example—Padron and Shishito peppers were little more than a rumor when I wrote the first edition of this book. I enjoyed them in Barcelona in the mid-1990s and 2000s but they were simply nowhere to be found in the United States. And then, one by one, farmers began responding to requests from customers. Now, they are almost everywhere, in farmers markets and grocery stores, on restaurant menus and on dinner tables.

Padrons have the nickname "Russian Roulette Peppers," as typically one in ten or so is fiery. But left on the vine, they will all become hot. If you don't care for such heat, your best bet is to choose the smallest Padrons you can find. Shishitos are rarely if ever hot.

1 pound Padron chiles, Shishito chiles, or a mix of both
Olive oil
1 lemon, halved
1 piece burrata, 4 to 6 ounces
Maldon Salt Flakes or Sel Gris
Black Pepper in a mill

Set a heavy pan, such as a cast-iron skillet, over medium-high heat, pour in a little olive oil (enough to coat the pan), and when the pan is hot, add the chiles. Fry, tossing frequently, until they just begin to soften and take on a bit of color. Remove from the heat.

Working quickly, put the burrata in the center of a plate.

Cut the lemon in half and squeeze the juice of one half over the chiles. Toss and add to the plate, letting them fall randomly. Cut the second half of the lemon in small wedges and add to the plate.

Season all over with salt and pepper and serve immediately.

Edamame
Fresh Soy Beans with Sea Salt

Serves 4 to 6

Edamame, once seen in America only in Japanese restaurants, is now so popular that you can find commercial versions in produce departments, in delis, and in freezer compartments. It is best, of course, when you make it at home, which is very easy to do.

1 pound fresh soybeans in their pods
1 tablespoon coarse sea salt, and more to finish

Toss the soybeans with the salt and let them sit for 15 minutes. Bring 2 quarts of water to a boil, add the soybeans, return the water to a boil, and simmer for 2 to 3 minutes. Drain thoroughly, place the beans in a bowl, season with a little more salt, and serve hot or at room temperature, with a bowl alongside for the discarded pods.

To eat the soybeans, pinch off one end of the pod and squeeze the beans into your mouth.

Deviled Eggs with Smoked Salt & Green Peppercorns

Serves 4 to 6

It's important to know how to make perfect deviled eggs, a classic American appetizer and picnic dish. A purist, I eschew additions such as pickle relish, capers, minced onions, or olives, but think a strong jolt of aromatic black pepper is essential. One delicious variation includes green peppercorns with black, and one is made with an intoxicating smoked salt.

6 large farm eggs
⅓ to ½ cup mayonnaise
3 tablespoons Dijon mustard or Green Peppercorn Dijon
1 teaspoon black peppercorns, crushed
1 teaspoon brined green peppercorns, drained and crushed
¾ teaspoon smoked salt, lightly crushed, plus more as needed

Put the eggs in a medium saucepan and cover with water by at least 1 inch. Set over medium heat, bring to a boil, simmer gently for 3 to 4 minutes, cover, remove from the heat and let rest, covered, for 15 minutes. Transfer the eggs to a cold-water bath, changing the water two or three times as it warms. Add ice to the water, crack each egg all over and let it rest in the ice water for another 15 minutes. Carefully peel the cooled eggs, rinse off any bits of shell, dry them on a tea towel and cut them in half lengthwise. Scoop out the yolks and leave the whites on your work surface.

Press the yolks through a potato ricer into a bowl or put them in the bowl and mash them with a fork. Mix in ⅓ cup of the mayonnaise, the mustard, peppers, and salt. If the mixture seems a little dry, add the remaining mayonnaise. Taste and correct for salt. Fill the centers of the egg whites with the egg yolk mixture, cover gently with wax paper or plastic wrap, and refrigerate for at least 30 minutes before serving.

Hard-Cooked Farm Eggs with Smoked Salt & Butter

Serves 6 as an appetizer, 3 as a main course

A decade ago, it would have seemed odd to suggest that a hard-cooked egg could make a suitable appetizer or main course. But our eggs have gotten so much better, as we have become, at least to some degree, more compassionate in how we raise chickens and more aware of the reality that their diet is inseparable from the quality of their eggs. Now, farm eggs are offered on high-end restaurant menus and at cocktail parties. At the same time our eggs have soared in quality, smoked salt has become more readily available. When I wrote the first edition of this book, I had tasted it just once and barely had time to slip it into the manuscript. Then I felt that eggs made the very best canvas for this delicious salt. All these years and many experiments later, I still feel that way. If you prefer soft-cooked eggs, consult the Variation at the end of this recipe.

6 farm eggs, preferably organic, from pastured hens
2 tablespoons butter, preferably organic, at room temperature
Smoked salt, lightly crushed
Black pepper in a mill
Butter toast points, if serving as a main course

Put the eggs in a medium saucepan, cover with water by at least 1 inch, and bring to a boil over medium heat. Simmer for about a minute, cover the pan, remove from the heat and let rest for 15 to 20 minutes.

Working quickly, set individual bowls or cups on your work surface.

Hold a folded tea towel in your non-dominant hand and, with a good teaspoon in your dominant hand, smack the egg hard at its equator so that the spoon goes all the way through it. Scoop out the egg, put it in the vessel and add a teaspoon of butter.

When all of the eggs are in their containers, season with salt and pepper and serve immediately.

Variation:

For soft-cooked eggs, simmer the eggs very gently for 2 minutes (for small eggs) or 3 minutes (for larger eggs), remove from the heat, and immediately take the eggs out of the water. Continue as above, protecting your hand with several folds of a tea towel. Omit the butter, season each egg with salt and pepper, and serve hot.

Bruschetta with Sautéed Greens

Serves 4

"You will see how it tastes of salt, the others' bread," Dante, one of the world's most famous Tuscans, wrote in The Divine Comedy. *In his newsletter "The Art of Eating," Edward Behr describes—which is not to say explains, no one really knows why—the tradition of unsalted bread in Tuscany, of the Tuscans' expectation that bread dough not be salted. Inevitably salt appears, of course. In the simplest version of bruschetta, called fettunta, bread is toasted, drizzled with freshly pressed olive oil, and topped with salt. Although not the traditional salt of Italy, sel gris has a robust flavor that is excellent with any type of bruschetta: with just olive oil and garlic, with diced tomatoes, or with sautéed greens, perfect in the spring before tomatoes are in season. In the summer, combine the two (see Variation).*

3 tablespoons olive oil
¼ pound sturdy greens (such as chard, Lacinato kale, beet greens, radish greens), rinsed
3 to 4 garlic cloves, pressed
Crushed red pepper flakes

Kosher salt
8 slices country-style bread
3 tablespoon extra virgin olive oil
Sel gris
Black pepper in a mill

Heat a stove-top grill or prepare a fire in a charcoal grill. When the grill is ready, prepare the greens. Heat the olive oil in a sauté pan set over medium heat, add the greens and cook for 3 minutes. Use tongs to turn the greens and continue to cook, turning now and then, until wilted and tender; time will vary depending on the variety of greens.

Add the pressed garlic and a generous shake or pinch of red pepper flakes, season lightly with kosher salt, remove from the heat, cover and keep hot.

Toast the bread on the grill until it is golden brown. Place the bread on a platter, drizzle each piece with about a teaspoon of extra virgin olive oil, and top with some of the greens. Sprinkle a little sel gris over each piece, grind black pepper over all, and serve immediately.

Pineapple with Black Pepper & Hawaiian Alaea Salt

Serves 8 to 10

A few wedges of pineapple make a wonderful appetizer either before a rich main course, such as Kalua Pig (page 218) or Black Pepper Crab (page 205), or on a very hot night. Salt adds a bright dimension, a spark, to the pineapple and black pepper adds a sultry flavor and aroma. The challenge in this simple recipe is in finding excellent pineapple, no easy task unless you're in an area where pineapples are grown. The pineapples sold in Hawaii, for example, are vastly better—more tender, less tart, and much sweeter—than those that are shipped to the mainland. When purchasing pineapples in mainland markets, look for those that have a fruity aroma, especially near their cut end, which should have a bit of give when pressed.

1 large ripe pineapple
Hawaiian alaea salt
1 tablespoon black peppercorns, coarsely crushed

Place the pineapple horizontally on your work surface. Use a heavy, sharp knife to cut it in half lengthwise, cutting through the crown and stem as well as the fruit itself.

Cut each half of the pineapple in half again, lengthwise. Use a paring knife to cut out the core at the top of each quarter, inserting the knife at one end and pulling it along to the other. Lift off and discard the core.

Using a sharp, somewhat flexible knife, cut each quarter wedge of flesh away from the skin; leave it in place. Cut each wedge into slices about ¾ inch thick, being sure not to cut through the skin.

Set the wedges in their skins on a serving platter. Using your fingers and holding your hand several inches above each wedge, shower the pineapple lightly with salt, followed by a heavier sprinkling of black pepper. Cover lightly with plastic wrap and chill for 30 minutes before serving.

Onandaga Salt Potatoes

Serves 8 to 10 as part of an appetizer buffet

Sometimes I calls this dish "funeral potatoes," as I have taken them to many memorial services, where they typically vanish quickly. They are so irresistibly delicious that they eclipse the diminished appetite of those in grief. They have an interesting history, too, and now and then I've encountered someone from Onandaga County in New York who recognizes them.

I first learned of these potatoes in an email message that arrived one afternoon while I worked on the first edition of this book. "Don't forget our 'salt potatoes,'" Valerie Jackson Bell, curator at the Salt Museum in Onandaga County, wrote.

According to local folklore, back when the salt factory was in operation, kids would toss small white potatoes into the pots of boiling brine, a saturated brew about fifty times as salty as seawater, fish them out, and enjoy them on the spot.

Locals still cook potatoes in the same way, though on top of the stove and not in an outdoor cauldron, she added, and serve them dripping with melted butter. You will be amazed just how flavorful the potatoes become in the brine. For the best results, choose tiny potatoes and don't skimp on the butter. A saturated brine is 26.4 percent salt, which equals 2.64 pounds of salt per gallon of water. The water in this recipe is not saturated, but there is enough salt in it to flavor the potatoes.

1½ cups kosher salt or unrefined sea salt
2 pounds very small new potatoes
6 tablespoons unsalted butter
Black pepper in a mill

Fill a medium pot with 2 quarts water, add the salt, and set over high heat. Stir to dissolve the salt, add the potatoes, and bring to a boil. Reduce the heat to medium-low and simmer the potatoes until they are completely tender when pierced with a fork, about 20 or 25 minutes, depending on the size of the potatoes.

Drain thoroughly, put in a serving bowl, add the butter, and toss lightly until the butter is melted. Season generously with black pepper and serve hot.

Grilled Figs with Prosciutto & Black Pepper

Makes 20 pieces

Sometime in mid-summer, figs begin to ripen and then there is a frenzy of them, begging for our attention. The late Jack McCarley of Green Man Farms in Healdsburg, California, shares his favorite way of enjoying the figs he grew on his farm. They make an excellent appetizer but are also a good side dish with roasted chicken and grilled duck breast. To serve as a first course, set the figs on a bed of fresh greens, and drizzle with a little sherry vinegar or lemon juice before seasoning with black pepper.

10 black Mission figs, firm-ripe
10 slices (about ¼ pound) prosciutto, cut in half crosswise
20 bamboo skewers, soaked in water for 1 hour
Black pepper in a mill

Prepare a fire in a charcoal grill at least 1 hour before grilling the figs.

Cut the figs in half lengthwise, wrap each half in a slice of prosciutto, and secure it with a bamboo skewer, breaking off any extra length of skewer and discarding it. Grill the figs for 3 or 4 minutes on each side, transfer to a platter, grind black pepper over them, and serve immediately.

Grilled BLT Kabobs

Makes 10 Kabobs

One of the simplest yet most compelling combinations of flavors is salty bacon, sweet tomatoes, and creamy mayonnaise, the tastes that transform a simple sandwich into the magical BLT, arguably American's finest sandwich. This recipe is adapted from one in "The BLT Cookbook," (Morrow, 2003), written as my love letter to this combination of flavors.

Thirty 1 ½-inch cubes best-quality sourdough hearth bread
⅓ cup extra virgin olive oil
Kosher salt
Black pepper in a mill
10 slices lean bacon
Ten 12-inch-long wooden skewers, soaked in water for 1 hour
30 firm-ripe cherry tomatoes
4 to 5 cups shredded Romaine lettuce
⅓ cup mayonnaise or aioli, preferably homemade, thinned with a little warm water or lemon juice, in a squeeze bottle

Put the bread into a medium bowl, drizzle the olive oil over it and toss until the bread has absorbed all the oil. Season with salt and pepper and toss again. Set aside.

Fry the bacon until it is not quite crisp and still completely pliant; drain on absorbent paper.

To make the kabobs, put a skewer through a slice of bacon, spearing it through the lean part near one end. Push it down about 1 ½ inches from the tip of the skewer. Add a bread cube and then pierce the bacon again, so that it is folded over the bread. Add a cherry tomato and fold the bacon over that. Continue until you have 3 cubes of bread and 3 sherry tomatoes, each separated by a fold of the bacon, on each skewer.

Grill over a medium charcoal fire or on a stove-top grill, turning to grill the bread on each side, for about 3 minutes.

Working quickly, toss the lettuce with a little salt and spread over a large serving platter. Set the grilled kabobs on top of the lettuce and squeeze a little mayonnaise or aioli over each one.

Serve immediately.

Taramosalata

Serves 6 to 8

    Caviar probably best demonstrates the ability of salt to transform the edible into the delectable. From highly prized osetra and beluga caviars to humble salmon and mullet roes, it is salt that makes a culinary delicacy out of a sack of fish eggs. In this traditional Greek dish, dried bread and olive oil soften an intensely salty roe (traditionally, mullet or carp), creating a delicious condiment that perks up the palate and stimulates the appetite. I prefer a version made with potatoes, also traditional though not as common.

One 7-ounce jar taramá (mullet roe) or other inexpensive caviar
Half a small red onion, minced (about ¼ cup)
2 medium russet potatoes, baked and riced (see Note)
1 tablespoon minced lemon zest
Juice of 2 lemons
⅔ cup extra virgin olive oil
2 tablespoons minced fresh flat-leaf parsley
1 baguette, thinly sliced
2 bunches small radishes, cleaned and trimmed
1 cucumber, thinly sliced
½ cup olives of choice

Using a large suribachi, grind together the taramá and onion until fairly smooth. Use a rubber spatula to fold the potato purée and the lemon zest into the mixture. Add the lemon juice slowly, mix until completely incorporated, and then add the olive oil, a bit at a time, mixing well after each addition. Add the parsley, taste, and add a little more olive oil if the mixture is too tart. Cover with plastic wrap and chill for at least 1 hour before serving.

Preheat the oven to 300 degrees. Place the baguette slices on a baking sheet, brush with a little olive oil, and toast until golden brown, about 15 minutes. To serve, place the toasts in a small basket. Transfer the taramosalata to a serving bowl, set it on a plate, and surround with the radishes, cucumber slices, and olives.

Serve immediately, with the basket of croutons alongside.

NOTE

Pierce scrubbed, unpeeled potatoes in several places with a fork and bake at 375 degrees until tender, about 40 minutes. Let the potatoes cool until they are easy to handle. Use both hands to break each potato in half and then place the halves, one at a time, into a potato ricer with the interior part of the potato facing into the ricer. Press the potatoes through (as you would garlic through a press), open the ricer, and remove and discard the skin. Repeat. There it no substitute for the light, uniform texture of riced potatoes; mashing results in a more dense and less even mixture.

Roasted Marrow Bones with Artichoke & Green Olive Tapenade

Serves 4

    Bone marrow, preferably from a grass-fed animal, is a delicious treat, luscious and rich on the palate and full of essential nutrients, some that we get from few other sources. Paired with a tangy condiment or salad (see Variation) and a glass of sparkling wine, roasted bone marrow is an elegant appetizer and even a perfect main course if followed by a big green salad. For appetizers, allow one bone per person; for a main course, allow two. If you can get marrow bones that are split lengthwise, do so. My neighborhood restaurant, K & L Bistro, serves their bones split and it takes much left effort and is less messy than when the bones are whole.

Artichoke & Olive Tapenade, page 353
Black pepper in a mill
4 veal or beef shank bones, 3 to 4 inches long, meat removed
4 to 6 slices grilled hearth bread
Maldon Salt Flakes

First, set a heavy pan (a cast-iron skillet is ideal) in the oven and preheat to 450 degrees.

While the oven heats, make the tapenade and set it aside.

Carefully set the bones in the hot pan, placing them vertically, not horizontally. Roast for about 15 to 20 minutes, depending on the size of the bones. Check after about 12 minutes to make sure that the marrow is not melting into the pan.

Moments before the bones are done, divide the tapenade between two plates. Add the bones and the toast, add a little pile of salt to each plate and serve immediately, preferably with a marrow spoon for each guest.

Red Miso Soup with Garlic, Ginger & Noodles

Serves 4

Miso is a thick fermented bean paste, sometimes salty, sometimes sweet, used to make the clear, delicate soup that is almost always served before sushi. It is a common breakfast dish in Japan, as well, because it is very high in protein, about 37 percent. Miso also is used to make many dressings and sauces.

There are several kinds of miso. White miso is mild, delicate, and sweet. Yellow miso, which includes rice mold, is salty and tart, and is the easiest to find in the United States, although other kinds are increasingly common. There are several types of red miso, but virtually all include rice and barley in addition to beans. Red miso is salty, full-flavored, and excellent for soup. Dark brown miso, generally made with just beans, is the most robustly favored; it is intensely salty.

4 cups Dashi (page 117)
2 slices fresh ginger
4 garlic cloves, sliced
½ teaspoon crushed red pepper
4 tablespoons red miso
2 tablespoons mirin (sweetened sake, for cooking)
3 ounces sömen (thin white wheat noodles), cooked and rinsed
3 scallions, white and light green part, trimmed and cut into thin rounds

In a soup pot, bring the dashi to a boil, add the ginger, garlic, and chili flakes, cover, and remove from the heat. Let steep for 15 minutes. Use a small strainer to remove and discard the ginger and garlic. Place the miso in a small bowl, add 3 or 4 tablespoons of the hot dashi, and stir until smooth. Return the dashi to the heat, stir in the miso and mirin, add the noodles, and when the soup is just hot, remove from the heat. Divide among 4 small soup bowls, using tongs to distribute the noodles evenly. Top each portion with some of the scallions and serve immediately.

Lemon Pepper Chicken Drumettes

Serves 6 to 8

If you don't want to go to the trouble of making the drummettes yourself (although it doesn't take long), you can usually purchase them at a good meat market. If you have homemade preserved lemons (see page 281 for a recipe) made with lemon juice (rather than brine), you can use some of the liquid from those lemons in this dish. Otherwise, use fresh lemon juice in which you've dissolved a tablespoon of kosher salt.

1 dozen chicken wings
1 tablespoon kosher salt
1 tablespoon freshly cracked black
    peppercorns
Generous pinch of dried oregano,
    optional

⅓ cup freshly squeezed lemon juice
2 or 3 lemons, very thinly sliced
2 tablespoons olive oil
Black pepper in a mill
Fleur de sel or sel gris for garnish

To make drumettes, cut the chicken wings at each joint. Use the two larger pieces for this recipe and reserve the wing tips for making stock.

Put the drumettes in a large bowl and toss with the tablespoon of salt and pepper (and the oregano, if using). Add the lemon juice and toss again; cover and refrigerate for at least an hour and as long as four hours.

To finish, remove the drumettes from the refrigerator.

Arrange the lemon slices, overlapping them slightly, over the surface of a seasoned clay baking dish just large enough to hold the wings in a single layer. If you don't have a seasoned clay dish, use a glass baking dish. Arrange the drumettes in a single layer on top of the lemons and add any juices that have collected in the bowl.

Put the dish in a cold oven, set the heat to 375 degrees and bake until tender and lightly browned, about 30 minutes.

Serve directly from the clay pot or transfer from a glass baking dish to a serving platter, sprinkle with a little fleur de sel or sel gris and serve immediately.

Brandade de Morue

Serves 6 to 8

Brandade de morue is a classic French bistro dish, generally served as an appetizer or first course, with toasted croutons alongside.

Depending on where in the United States you live, salt cod either is ubiquitous—as it is in Rhode Island, where the Italian and Portuguese communities keep the demand high—or very hard to find. Because of health department regulations in some states, California, for example, you won't see it hanging above deli cases, as it does in Europe. It will be refrigerated, often in a special compartment in the back, so if you don't see it, ask.

In order to leach out excessive salt, salt cod needs to be soaked in a water bath that is changed every few hours; I have found that twelve hours is usually sufficient; occasionally, a longer soak is a required. Should you be distracted, your cod will still be okay after three days in its bath. Once soaked, salt cod is no longer salty, and the finished dish will need to be properly seasoned before it is served.

1 pound boneless, skinless salt cod, soaked for 1 to 2 days
2 or 3 russet potatoes (about 2 pounds), baked until tender and riced while
    hot (see Note on page 107)
10 garlic cloves
1 teaspoon kosher salt
⅔ cup heavy cream
2 to 3 teaspoons black peppercorns, crushed
⅔ cup best-quality extra virgin olive oil
1 baguette, sliced and toasted

Remove the salt cod from its last cold-water bath, place it in a large sauté pan, cover with fresh cold water, and set over medium heat. When the water boils, remove the pan from the fire, and cover it. Let the cod sit for 15 minutes, drain it thoroughly, and cool. Break the fish into small pieces and pick out any bits of bone. Set aside.

Using a suribachi or a mortar and pestle, crush the garlic with the kosher salt until it is nearly liquefied.

Combine the cream and 2 teaspoons of the black peppercorns in a small saucepan and scald over medium heat. Set aside.

Put the salt cod in the container of a food processor, add the garlic, and pulse several times. With the machine running, pour in the olive oil in a slow drizzle, followed by the hot cream. When both are completely incorporated, transfer the mixture to a medium bowl. Use a heavy wooden spoon or wooden spatula to fold in the riced potatoes. Taste the brandade, correct for salt and fold in the remaining pepper.

Serve warm, with toasted baguette slices.

Soups

While signing books at a trade show, I was approached by a young woman who said she never made soup because hers did not turn out as well as her mother's. She explained that she used the best ingredients she could find to make stock, but still her soups were bland and flavorless.

When I inquired about salt, she looked horrified. "Salt? I never, ever use salt in anything," she said and walked away in a huff.

Soup without salt is bland. Garlic soup from Provence or Mexico tastes insipid, chilled avocado soup is flat, gazpacho dull. Japanese soups rely on dashi, a mildly salty seaweed stock, for flavor, but other soups require salting for both flavor and balance.

Pepper, too, is an important ingredient in most soups the world around. From restorative chicken broths to the rich soup topped with dumplings and known as Philadelphia Pepper Pot, pepper adds depth and character, completeness, you could say, to soup. Black pepper can serve to balance the sweetness of certain ingredients, too. Both salt and pepper play leading roles in many soups.

No matter what style of soups you prefer, expert use of salt and pepper is essential. This, of course, makes it a bit difficult to narrow the focus of this chapter and so I have featured soups that rely on salt, pepper, or both in ways beyond simple seasoning.

Dashi

Yogurt Gazpacho

Black Pepper Soup

Mulligatawny Soup

Salt Cod Chowder

Ginger Beef Noodle Soup

The Simplest Potato Soup

Spicy Sweet Potato Soup with Nutmeg Cream

Chilled Pear Soup

Dashi

Seaweed Stock

Makes 2 quarts

Sea vegetables concentrate the aroma of seawater, the essence of which is salt, into highly nutritious bundles of flavor, sometimes with such intensity that if you are unfamiliar with them you'll find them unpleasant. They take some getting used to, and a soup using a seaweed stock is a subtle way of introducing these good flavors to the novice. Dashi, a simple stock made of giant kelp and dried fish, is a critical building block of Japanese cuisine; you cannot achieve essential delicate flavors without it. It is remarkable that it's so easy to make. Because I live close to the Pacific Ocean, many types of seaweed are readily available, much of it gathered, dried, and sold by local vendors. There have been times when I craved some miso soup, had no konbu, but did have nori and dulse. I've made stock using them, and although the results are not absolutely traditional, I was pleased and fortified by my soup. You can find both seaweed and dried bonito in many health food stores and most Asian markets.

1 ½ ounces konbu (giant kelp)
1 ounce dried bonito flakes

Pour 2 quarts cold water into a large soup pot, add the konbu, heat the water very slowly, and just before it boils, turn off the heat and cover the pot. Let sit for 1 hour. Test the konbu by pinching it with a fingernail; if it is tender, remove it from the pot; if it is still tough, let it sit for another hour.

Add the bonito flakes to the konbu stock, slowly bring the liquid to a boil, and remove from the heat the moment it boils. Let the pot sit without moving it for about a minute as the flakes settle to the bottom. Skim off any foam that has formed, and then strain the stock through 3 or 4 layers of cheesecloth; save the bonito flakes for secondary stock. Use immediately, or cool to room temperature and store, covered, in the refrigerator for up to 4 days.

Yogurt Gazpacho

Serves 3 to 4

If you want to show off this soup in the most dazzling context possible, invest in either a large Himalayan salt bowl or several individual salt bowls. They are pricey and will, eventually, vanish, as foods in them absorb the salt. Mostly, they are a pretty indulgence, a grown-up plaything. If cost is not an issue, they are great fun. This soup is a perfect use, as it will blossom from its contact with the salt.

But if you don't have or don't want bowls made of salt, the soup will be beautiful in any pretty vessel. And no matter the vessel, it will be delicious.

This soup improves for a couple of days, so don't be intimidated by the quantity. It won't go to waste, I promise.

2 pounds ripe tomatoes, peeled, halved, and seeded
Kosher salt
2 cups whole milk yogurt
1 serrano, minced
1 cucumber, peeled, seeded, and minced
Juice of ½ lemon
1 tablespoon medium acid red wine vinegar, such as B. R. Cohn Cabernet
    Sauvignon Vinegar
Black pepper in a mill
2 tablespoons minced fresh cilantro
Best-quality extra virgin olive oil
2 teaspoons za'atar (see Note on next page)

Set a strainer over a large deep bowl.

Use a very sharp knife to mince the tomatoes, scooping up the pulp and the juice frequently and transferring it to the strainer. Continue until all of the tomatoes have been minced.

Stir 2 teaspoons of salt into the tomatoes and set aside for several minutes.

Put the yogurt into a large soup bowl, add the serranos, cucumbers, lemon juice, and vinegar and stir.

Fold in the drained tomato pulp. Reserve the drained juices (chill and enjoy neat or in a Bloody Mary).

Taste and season generously with salt, tasting after each addition. When the flavors blossom, you have added enough. Season with black pepper and stir in the cilantro.

Chill thoroughly, for at least two hours and as long as overnight.

To serve, ladle the soup into chilled soup plates. Drizzle each portion with a little olive oil and sprinkle za'atar on top. Serve immediately.

NOTE

Za'atar is a Middle Eastern spice mixture, a combination of of sumac, toasted and lightly crushed sesame seeds, dried thyme, and salt. Some versions include hyssop. The sumac contributes both a deep red color and a lemony tang. It is available online at such locations as www.thespicehouse.com.

Black Pepper Soup

Serves 4 to 6

San Francisco's Mandalay Restaurant on California Street, the city's first Burmese restaurant, serves a wonderfully fragrant black pepper soup. There's more black pepper in the soup than you'd think you could add successfully, yet the soup is not overwhelming or out of balance. After the soup has been eaten, a film of black pepper glistens in the bottom of the bowl, a testament to this robust and unusual dish.

1 bunch (about 10 ounces) fresh Swiss chard, thoroughly rinsed

2 ounces dried Chinese black mushrooms, soaked in ¾ cup water for 30 minutes

8 cups homemade chicken broth (see Note below)

1 ¼-inch thick slice of fresh ginger

2 tablespoons freshly crushed black pepper

2 zucchini, sliced into thin rounds

1 pound rock fish fillets, boned, cut into 1-inch pieces and lightly salted

Kosher salt

4 ounces rice vermicelli or other very thin noodle, softened in hot water and drained

¼ cup fresh cilantro leaves

Trim the stems from the chard and reserve them for another purpose. Cut the leaves crosswise into ½-inch slices. Set aside.

Strain the liquid from the mushrooms through a coffee filter or very fine strainer and combine it with the chicken broth, ginger, and black pepper in a large soup pot. Set over medium heat, bring to a boil, reduce the heat, and simmer for 10 minutes.

Add the soaked mushrooms, zucchini, rock fish, and chard, and simmer until the fish is cooked through, about 5 minutes. Taste and correct for salt, if needed.

Add the vermicelli, heat through, and ladle into large soup bowls. Top each portion with some of the cilantro leaves and serve immediately.

Mulligatawny Soup

Serves 4 to 6

Tamil Nadu, a state in southern India, is the home of Piper nigrum. *Tamil is the language of the native population, also known as Tamil, and it is their expression for "pepper water,"* molaga-tanni, *that gives us the name of this classic English soup, Mulligatawny. As the story goes, a British colonist requested some more of that "mulligatawny," referring to a spicy soup he'd enjoyed. Countless versions have been made since that first awkward request, though the main characteristics—a mild spiciness and heat, a bit of sweetness from apples, and chicken—are constants. In this version, I've used the chicken-apple sausages that have become so popular in recent years and finished the soup with tender sautéed spiced apples.*

1 cup raw jasmine rice
2 pounds chicken-apple sausages
1 cup fruity white wine
2 firm sweet-tart apples, peeled and cut into ⅛-inch slices
2 teaspoons curry powder, commercial or homemade
½ teaspoon ground cumin
Black pepper in a mill
One 2-inch piece cinnamon
1 yellow onion, minced
2 carrots, peeled and minced
4 garlic cloves, minced
2 teaspoons grated ginger
2 teaspoons kosher salt
1 teaspoon turmeric
¼ teaspoon cayenne pepper
4 cups chicken stock
One 14-ounce can coconut milk

Put the rice in a strainer or colander and rinse it under cool running water until the water runs clear. Place the rice in a large saucepan with 1 ½ cups cold water and bring to a boil over high heat. Reduce the heat to low, cover the pan, and cook for 20 minutes without lifting the lid. Remove from the heat and let the rice steam undisturbed for 10 minutes before fluffing with a fork. Fluff the rice, cover, and set aside while you prepare the soup.

Prick the sausages in several places with a fork, put them in a sautè pan, add the wine, cover, set over medium heat, and simmer for 7 or 8 minutes. Remove the lid, increase the heat, and brown the sausages evenly all over. Set aside and let cool. Cut into ¼-inch slices.

Cut the slices of apple in half crosswise. Melt 2 tablespoons of the clarified butter in a large soup pot, add the apples, and season with a pinch of the curry powder, a pinch of the cumin, and several turns of black pepper. Add the cinnamon stick. Sauté the apples until they are tender and golden brown. Remove from the pot, leaving the cinnamon stick behind, and set aside.

Heat the remaining 2 tablespoons of the clarified butter in the same pot, add the onion and carrots, and sauté over medium-low heat until very soft and fragrant, about 15 minutes. Add the garlic and ginger, and sauté for 2 minutes more. Stir in the curry powder, cumin, salt, turmeric, and cayenne, add the chicken stock and 3 cups of water, bring to a boil, reduce to a simmer, and cook for 15 minutes. Add the sausages and simmer 5 minutes more. Stir in the coconut milk, add 1 teaspoon freshly ground pepper, taste, and correct the seasoning. Divide the rice among individual soup bowls, ladle the soup over the rice, top with the apples, and serve immediately.

Salt Cod Chowder

Serves 4 to 6

One morning I received a phone call from a woman in her seventies who reads my weekly columns in the Santa Rosa Press Democrat *and wanted to share recipes from her mother, who immigrated from Portugal in the early 1900s. One of the recipes was for a hearty soup of fresh fava beans, rock fish, and linguiça, which I've used for inspiration for this Portuguese-style chowder. Linguiça is a spicy pork sausage seasoned liberally with garlic, cayenne pepper, black pepper, and a small amount of red wine, which contributes a characteristic acidity. It is widely available in the United States, but if you can't find it, use kielbasa, andouille, Spanish chorizo, or another spicy sausage instead. If you have Portuguese salt, which currently is being imported by a few companies in the United States, be sure to use it here for a flourish of authenticity.*

½ pound salt cod, soaked for 1 to 2 days (see Brandade de Morue, page 112)
1 pound linguiça, or other spicy pork sausage cut into ¼-inch rounds
1 onion, diced
6 garlic cloves, minced
Kosher salt
1 pound waxy potatoes, such as Yellow Finn or Yukon Gold, diced
One 28-ounce can whole tomatoes, with juice (preferably Muir Glen brand)
6 cups chicken stock
3 ounces (¾ cup) cracked green olives, pitted and sliced
¼ cup dry Madeira
Black pepper in a mill
8 thick slices country-style bread, toasted and rubbed with garlic
Olive oil

Remove the salt cod from its final cold-water bath, place it in a large sauté pan, cover with fresh cold water, and set over medium heat. When the water boils, remove the pan from the fire, and cover it. Let the cod sit for 15 minutes, drain thoroughly, and cool. Break the fish into small pieces and pick out any pieces of bone. Set aside.

Fry the linguiça in a large soup pot set over medium heat until most of the fat has been released. Using a slotted spoon, remove the linguiça from the pot and set it aside. Fry the onion in the fat until it is limp and fragrant, about 7 to 8 minutes. Add the garlic and sauté for 2 minutes more. Season with a little salt and return the linguiça to the pot, add the salt cod, potatoes, tomatoes, and stock, bring to a boil over high heat, reduce the heat, and simmer, covered, for 25 minutes, until the potatoes are completely tender.

Add the olives and Madeira, and simmer for 5 minutes more. Taste and season with salt and pepper.

Place a piece of toast in each of 4 individual soup plates, drizzle with olive oil, and then ladle the soup over the bread. Serve immediately, with the remaining toast alongside.

Variation:

To make this soup with fresh fish, omit the salt cod. Cut 1 pound of rock fish into 1 ½-inch pieces, season them lightly with salt and pepper, add them to the soup with the olives and Madeira and be certain not to overcook. Taste and correct for salt.

Ginger Beef Noodle Soup

Serves 4 to 6

The night I arrived in Kuching, Sarawak, on the island of Borneo, I went down to the recently renovated river walkway that each night teems with young people, tourists, and hawkers offering delicious local and American foods. The very last stall had huge bowls of soup, served in the hot clay pots in which it was cooked to order. The soup had big chunks of beef and the broth was redolent of ginger, garlic, black pepper, and spicy chiles. There, under a tropical moon in the heavy night air, nothing had ever tasted better. I was so entranced that I forgot the time and had to run the half mile or so back to my hotel, where guides were waiting to take me to the Night Market on the other side of town.

1 quart fragrant beef stock (recipe follows), hot
3 tablespoons clarified butter
1 large yellow onion, diced
6 garlic cloves, minced
13-inch piece fresh ginger, grated
2 beef shanks, each about 2 inches thick
Kosher salt
2 stalks lemongrass, bulb parts only, crushed and very thinly sliced
1 teaspoon whole white peppercorns
1 teaspoon whole black peppercorns
1 teaspoon red pepper flakes
10 ounces rice noodles, ¾-inch wide
¼ cup fresh cilantro leaves
1 or 2 limes, cut into wedges

Make the stock at least one day before preparing the soup and heat it when you are ready to finish the soup.

To make the soup, melt the clarified butter in a wide soup pot set over medium-low heat, add the diced onion, and sauté until it begins to caramelize, about 25 minutes. Add the garlic and grated ginger, and sauté for 2 minutes more. Add the reserved 2 beef shanks and brown on all sides.

When the shanks are browned, season with salt and add the stock to the pan, along with 2 cups of water, the lemongrass, the white and black peppercorns, and red pepper flakes. Bring to a boil over high heat, reduce to low, and simmer, partially covered, until the meat is very tender, about 2 hours.

Prepare the rice noodles according to package directions and drain.

Remove the cooked shanks from the pot, chop the meat coarsely, and return it to the pot. Taste the broth, and correct the seasoning. Divide the noodles among individual soup bowls, ladle the soup over the noodles, garnish each portion with some cilantro leaves and a squeeze of lime, and serve immediately.

Spicy Beef Stock

Makes 1 quart

3 ½ pounds beef shanks, cut 2 inches thick through the bone
1 carrot, cut into 3 or 4 pieces
1 tomato, cut in half
1 yellow onion, cut in quarters
Kosher salt
3 stalks lemongrass
6 large slices ginger
3 kaffir lime leaves
2 star anise

To make the stock, preheat the oven to 375 degrees. Put the beef shanks, the carrot, the tomato, and the quartered onion into a large roasting pan, season fairly generously with salt and roast for 45 minutes.

Remove from the oven and transfer to a large soup pot, add 12 cups of water, the lemon grass, the sliced ginger, the kaffir lime leaves, and the star anise, bring to a boil over high heat, reduce to low, and simmer, partially covered, for about 4 to 6 hours, until the meat completely falls off the bone and the liquid is reduced by two thirds. Skim off any foam that collects on top of the liquid.

Let cool, strain, and refrigerate overnight. Remove and discard the layer of fat that forms on top.

The Simplest Potato Soup

Serves 6 to 8

When this soup is made with really good potatoes, it has an extraordinarily earthy flavor that is delicious simply as it is or as a canvas for many other flavors, such as sautéed wild mushrooms. When the black pepper is really good, too, a sort of alchemy takes place that infuses the earthy potatoes with rich, deep flavors. Add condiments, if you like, to resonate what will come next. A bit of caraway butter atop each serving is delicious when a rich beef stew or some sort of cabbage will follow. Add the zest of 2 lemons along with the black pepper if fish or shellfish is on the menu. You get the idea, right?

3 tablespoons butter
1 yellow onion, cut into small dice
Flake salt
2 ½ pounds organic dry-farmed
potatoes, such as German Butter
or Yukon Gold
6 to 8 cups homemade chicken stock
Black pepper in a mill

Put the butter into a large saucepan or soup pot, set over medium-low heat and, when the butter is melted, add the onion. Cook gently until very soft and fragrant, about 20 to 25 minutes; do not let the onion brown. Season with salt.

While the onion cooks, peel the potatoes and cut them into very thin slices.

Add the sliced potatoes to the pot, stir and cook for 2 minutes. Season generously with salt, add the stock and increase the heat to high. If necessary, add enough water to cover the potatoes by about 1 inch. When the liquid boils, reduce the heat and simmer very gently until the potatoes are tender, about 12 to 15 minutes.

Test the potatoes and if they are quite tender, remove from the heat. If they are not yet tender, cook another few minutes, test again and remove from the heat. Season very generously with black pepper, cover and let rest 10 to 15 minutes.

Use an immersion blender to purèe the soup until quite smooth. If it is too thick, thin with a little stock or water.

Return to a low burner, heat through, taste and correct for salt and pepper.

Spicy Sweet Potato Soup with Nutmeg Cream

Serves 4 to 6

If you find sweet potatoes almost too sweet, as I do, you'll love this recipe, which uses spicy and tangy elements—garlic, smoked chiles, vinegar, and lots of black peppercorns—to mitigate the sweetness of the potatoes and create a deliciously complex yet easily prepared soup. This is an excellent way to begin a fall or winter meal.

3 tablespoons olive oil
1 yellow onion, diced
1 shallot, minced
6 garlic cloves, minced
½ teaspoon chipotle powder
2 teaspoons kosher salt, plus more to taste
2 teaspoons crushed black peppercorns, plus more to taste
3 medium sweet potatoes (about 2 ½ pounds total weight), peeled and
  thinly sliced
1 medium russet potato, scrubbed and thinly sliced
¼ cup apple cider vinegar
1 ½ cups apple apple cider
Nutmeg, whole
Nutmeg Cream (see Note on next page)
¼ cup chopped cilantro

Heat the olive oil in a heavy soup pot over medium heat, add the onion, and sauté until soft and fragrant, about 7 to 8 minutes. Add the shallot and sauté 5 minutes more, stirring occasionally to be sure neither the onion nor the shallot browns. Add the garlic and chipotle powder, stir, and sauté 2 minutes more.

Season with about a teaspoon of kosher salt and the 2 teaspoons of black pepper. Add the sweet potatoes, russet potato, vinegar, apple cider, and 3 cups of water. Increase the heat to medium, bring to a boil, reduce the heat, and simmer until the potatoes are tender, about 20 minutes. Add more water if necessary.

Remove from the heat and let cool slightly.

Use an immersion blender to purée the soup or pass it through a food mill. Add several gratings or grindings of nutmeg (about ½ teaspoon total), along with the remaining teaspoon of salt. Taste and correct for salt and pepper. Ladle into soup bowls, drizzle with a little nutmeg cream, top with cilantro and serve immediately.

NOTE

To make Nutmeg Cream, combine ¼ cup creme fraiche, 2 to 3 table-spoons half-and-half, ½ teaspoon freshly grated nutmeg, ½ teaspoon kosher salt, several turns of black pepper from a mill set on medium-coarse, and a pinch of sugar in a small bowl. Taste, correct seasoning, and refrigerate, covered, until ready to use.

Chilled Pear Soup

Serves 4

This delicate and fragrant soup works beautifully as a first course and as a dessert, especially following a rich, spicy meal, where it is welcomingly refreshing. The pepper-corns are not overpowering, but rather add a sultry, sexy dimension that, when combined with the salty cheese and delicate pears, creates a compelling medley of flavors. Because the flavors in this soup are pure and pristine, use spring water if your tap water has off-flavors.

4 ripe Bartlett pears, peeled, cored, and sliced
2 slices fresh ginger
2 teaspoons whole black peppercorns
1 teaspoon whole white peppercorns
2 cardamom seeds (not pods)
Tiny pinch of salt
1 cup dry fruity white wine, chilled
2 ounces blue cheese, such as Point Reyes Original Blue
2 tablespoons fresh mint leaves, cut into thin strips

Place the pears, ginger, peppercorns, and cardamom seeds in a saucepan, and add 2 ½ cups water. Bring to a boil over medium heat, reduce the heat to low, and simmer until the pears are tender, about 12 minutes. Remove from the heat and cool the pears in the liquid.

Remove and discard the ginger slices. Use a slotted spoon to remove the pears from the cooking liquid; pass them through a fine sieve or food mill fitted with the smallest blade. Strain the cooking water, discard the peppers and seeds, and stir it into the pear purèe. Chill the mixture for at least 3 hours or overnight.

To serve, stir the wine into the pear purée.

Divide the cheese among four soup plates and ladle the soup over it. Garnish with a shower of mint leaves and serve immediately.

Salads

In Italy, salad is an expression of saltiness, literally. *Insalata,* which means salad in Italian, comes from the phrase *un'insalata,* that which is salted. A salad at its most basic is simply salted greens.

Most of us add salt after dressing a green salad, but I recommend trying the Italian method: Put fresh greens that have been rinsed and dried in a large bowl, sprinkle them with flake salt and toss them gently. Add a generous amount of extra virgin olive oil, toss again, and then toss them with a miserly amount of vinegar or citrus juice. I like to add several turns of black pepper, too, though it is not, strictly speaking, traditional. This type of salad should be made immediately before serving it, as should all salads made with leafy greens.

Today we use the word salad to mean all sorts of combinations of fresh and cooked foods. Usually, there is something acidic involved, such as vinegar or lemon juice, and there is almost always a fresh vegetable or fruit in there, too, along with all manner of grains, cheese, seafood, and meats. Salt and pepper, in one form or another, are essential for successful salads, regardless of their simplicity or complexity. Salt brings the flavors together, and often provides a pleasantly crunchy texture between the teeth, especially in salads of fresh tomatoes.

Pepper, too, contributes an essential and sultry quality to nearly all salads. It is essential, I believe, in fruit salads.

Other ingredients contribute both salty and peppery elements to salads. Some, such as radishes and their leaves, nasturtium greens, arugula, and watercress, are naturally peppery. Others, such as capers, olives, pickles, anchovies, and many cheeses, develop their flavors through brining, preserving in salt or the careful use of salt to both enhance and regulate the development of natural flavors.

Peppery Greens with Roquefort & Pears
Spicy Banana Raita
Summer Tomato Salad
Citrus Salad with Black Pepper
Citrus Salad with Feta, Hazelnuts & Arugula
Avocado, Grapefruit & Chicken Salad with Black Pepper Dressing
Golden Beet Salad with Feta & Garnet Vinaigrette
Watermelon, Feta & Arugula Salad
Chickpea, Celery, Green Olive, Scallion & Feta Salad
Carrot Salad with Black Olives, Pecans, Pomegranates & Feta
Poached Egg Salad with Frisee & Bacon Vinaigrette
Smashed Deviled Egg Salad with Chermoula & Spring Greens
Cobb Salad, Partially Deconstructed
Dungeness Crab Salad
Salt Cod Vinaigrette
Shrimp, Avocado & Arugula Salad with Radish Vinaigrette
Salt Cod, Potato & Artichoke Salad

Peppery Greens with Roquefort & Pears

Serves 4 to 6

Certain greens have a natural flavor that resembles black pepper. Some, such as nasturtium leaves, are fairly mild; others, such as peppercress, a relative of watercress, are bold and assertive. Arugula, too, depending on the variety and the season, can be full of peppery heat. The strongest should be used sparingly, in a mix with milder greens. Nasturtium flowers are both sweet and peppery. Many of the spicy greens, such as arugula, are best in cooler weather, when they are mild enough to be used on their own. Nasturtiums are abundant from late spring through the fall. Farmers markets are generally the best sources for specialty greens. If you have access to miner's lettuce, which is found almost exclusively in the wild, use it; it is delicately flavored and delicious.

2 cups (1 or 2 handfuls, about 1 bunch) peppery greens
4 cups (3 to 4 handfuls, about ⅛ pound) mixed young lettuces
¾ teaspoon flake salt
2 tablespoons walnut oil
1 tablespoon pear vinegar, lemon juice, or white wine vinegar
2 to 3 tablespoons mild extra virgin olive oil
Half a ripe pear, peeled, cored, and thinly sliced
2 ounces Roquefort cheese, crumbled
Black pepper in a mill

Put the greens and lettuces into a large salad bowl, sprinkle with the salt and toss gently. Add the walnut oil and turn gently until the greens are evenly coated. Sprinkle with the acid and toss again. Scatter the pears and cheese the top, transfer to individual plates, season with salt and several turns of black pepper and serve immediately.

Spicy Banana Raita

Serves 4 to 6

A raita is an Indian salad, nearly always made with yogurt, and generally served as one of several condiments to provide a contrast in temperature and texture. Their use is so common in this context that they are often identified as condiments or sauces. They are tangy and cool, a counterpoint to the dark richness of curried meats and vegetables.

Raitas may be made with diced tomatoes, cucumber and mint, potatoes and onions, chopped spinach, and sometimes apples, cauliflower, or bananas. All are mildly spicy, most often from cayenne pepper. I prefer the heat of fresh serranos and crushed black pepper, which provide a broader range of flavor than the one-dimensional cayenne does.

1 lime
1 or 2 serrano chiles
2 firm ripe bananas, peeled
2 teaspoons freshly crushed black pepper
3 cups plain, unflavored yogurt
2 tablespoons sugar
1 teaspoon kosher salt, or teaspoon each kosher salt and black salt (see glossary)
2 tablespoons minced fresh cilantro leaves
1 teaspoon cumin seed, roasted and coarsely crushed

Zest the lime and then juice it, setting aside the zest and the juice in separate containers.

Toast the serranos in a dry pan over medium-high heat until they are lightly browned and fragrant. Let them cool, remove the stem, and mince, removing the seeds first if you prefer a milder taste; set them aside. Slice the bananas ⅛ inch thick and toss with the lime juice and pepper; set them aside.

In a medium bowl, whisk together the yogurt, sugar, salt, black salt, lime zest, and minced serrano. Fold in the bananas and cilantro, taste, and correct the seasoning, scatter the cumin seed on top. Chill for 1 hour before serving.

Summer Tomato Salad

Serves 2 to 4, easily doubled

For a delicious tomato salad, you can get by with just two ingredients: perfectly ripe backyard tomatoes and good salt. You don't, of course, need a recipe. The variations on this theme are countless and you'll find many of them in The New Good Cook's Book of Tomatoes *(Skyhorse Publsihing, 2015).*

One way to change up the salad is to simply vary the salt you use. Flake salt will produce a juicy salad, as the grains extract the juices. Smoked salt adds its smoky flourish, fleur de sel or sel gris contributes an enticing element of crunch, and Hawaiian alaea salt adds a silky quality. Add freshly ground pepper—green, black, or white—and you have another layer of flavors to tickle your palate.

This salad is the one I make most often at the height of tomato season.

Use a vegetable peeler to make the curls of cheese.

3 to 4 ripe backyard-quality tomatoes, preferably an heirloom beefsteak
    variety, cored
Several very thin slices of red onion, rings separated
2 or 3 garlic cloves, minced
3 tablespoons chopped fresh Italian parsley or 8 to 10 leaves fresh basil,
    thinly shredded
Flake salt
Extra virgin olive oil
Black pepper in a mill
Several curls of cheese, such as Dry Jack or Pepato

Arrange the tomatoes in an overlapping circle on a pretty platter and tuck rings of onion here and there between them. Scatter the garlic and parsley, if using, over the top. If using the basil, set it aside momentarily.

Season all over with salt, drizzle with olive oil and add several generous turns of black pepper.

Scatter the curls of cheese on top and add the basil, if using.

Let the salad rest for about 5 minutes before serving.

Citrus Salad with Black Pepper

Serves 4 to 6

This salad was inspired by the delicious orange salads of Sicily, typically seasoned with salt, pepper, olive oil, and sometimes curls of cheese, and served before a rich meal. Here, I omit the olive oil as I find it eclipses the subtle harmony between the various fruits.

To peel citrus fruit, use a very sharp knife. First, cut each end flat, removing all of the rind to expose the fruit itself. Next, set the fruit on end and use the knife to cut from top to bottom, removing the peel as well as the rind beneath it in a single cut. As you peel, you will also be shaping the fruit. When each fruit is peeled, cut away any white pith you might have missed the first time around. Serve this salad on a large clear or colored glass platter so that you can see the beautiful juices of each fruit mingling.

3 grapefruit (Ruby Red or Star Ruby), peeled
2 blood oranges, peeled
3 oranges (Valencia or navel, but preferably both), peeled
1 lemon, preferably Meyer, peeled
1 lime, peeled
Flake salt
Black pepper in a mill

Cut the grapefruit into round slices (through the equator, not the poles) about ⅛ inch thick. Arrange them randomly on a glass platter in overlapping circles.

Season with a little kosher salt and several generous turns of black pepper, and set the salad aside for 30 minutes before serving so that the juices can mingle.

Before serving, add a very light sprinkling of salt.

Citrus Salad with Feta, Hazelnuts & Arugula

Serves 3 to 4

When you want something more than a green salad and tomatoes are not in season, citrus is a fabulous option and it is at its peak in the winter. Proper seasoning with salt and pepper highlights citrus fruits' subtle qualities, making them blossom deliciously.

2 grapefruit, peeled and sectioned
1 orange (preferably Cara Cara), peeled and sectioned
1 Meyer lemon, peeled and sectioned
Flake salt
Black pepper in a mill
3 generous handfuls small arugula
1 tablespoon hazelnut oil or toasted peanut oil
4 ounces brined feta, broken into chunks
¼ cup hazelnuts, lightly toasted

Put the grapefruit, orange, and lemon into a small bowl, season lightly with salt and a few turns of black pepper, toss and set aside briefly.

Put the arugula into a medium bowl, season lightly with salt, toss, drizzle the hazelnut or peanut oil on top and toss again. Divide among individual plates.

Spoon the citrus and its juices over the arugula and scatter the feta and hazelnuts on top. Season lightly with salt and a few turns of black pepper and serve immediately.

Avocado, Grapefruit & Chicken Salad with Black Pepper Dressing

Serves 4

Make this salad when you have roasted chicken left over from a recent meal. You can also use smoked chicken, available at many specialty stores and through mail order.

Black Pepper Dressing (page 357)
4 cups mixed peppery greens (watercress, peppercress, arugula, nasturtium
    leaves, radish leaves)
Flake salt
1 avocado, peeled and sliced lengthwise
1 pink grapefruit, peeled and sectioned
10 ounces roasted or smoked chicken, preferably thigh meat
1 lime, cut into wedges

Make the Black Pepper Dressing and set it aside.

Put the greens into a bowl, season with salt, toss and divide among 4 plates, placing them in a mound in the center of each plate.

Arrange a circle of avocado around each mound of greens and add sections of grapefruit next to some of the avocado slices. Put chicken on top of the greens. Spoon dressing over each salad, drizzling some on the chicken and more on the avocado slices and grapefruit sections.

Serve immediately.

Golden Beet Salad with Feta & Garnet Vinaigrette

Serves 3 to 4

Golden beets have a more delicate flavor than red beets and appeal to people who don't think they care for this root vegetable. I like to combine both, using the red beet sparingly in a pretty vinaigrette spooned over the golden beets. I find a generous amount of pepper perfectly balances a beet's sweetness.

6 to 8 medium golden beets, leaves removed, rinsed
1 small red beet, leaves removed, rinsed
Olive oil
1 garlic clove, minced
1 small shallot, minced
2 tablespoons best-quality red wine vinegar
Flake salt
Black pepper in a mill
6 tablespoons extra virgin olive oil
3 to 4 ounces feta, preferably sheep feta
1 teaspoon fresh thyme leaves
3 cups, loosely packed, very fresh small salad greens

Preheat the oven to 375 degrees.

Put the beets in a baking dish, drizzle with just enough olive oil to coat them lightly, set in the oven and cook until tender when pierced with a fork or bamboo skewer.

Remove from the oven and let cool until easy to handle.

Set aside the golden beets. Peel the red beet, using your fingers to pull off the skin. Cut it into very small dice and put it into a small bowl. Add the garlic, shallot, red wine vinegar, a very generous pinch or two of salt and several generous turns of black pepper, stir, set aside for 15 to 20 minutes and then whisk in the olive oil. Taste and correct for salt and black pepper. Set aside.

If there is beet juice on your hands, wash them thoroughly.

Peel the golden beets; the skins should come off easily. Cut each into 6 wedges and put into a medium bowl. Add the feta cheese and thyme leaves and toss gently. Add about 2 tablespoons of the vinaigrette and toss very gently.

Put the greens into a bowl, season with salt, toss and divide among individual plates; top with some of the golden beets and cheese. Drizzle a generous spoonful of dressing over each portion and serve immediately.

Watermelon, Feta & Arugula Salad

Serves 4

Watermelon salads have become extremely popular in recent years and for good reason: They are delicious. I find the best ones are typically the simplest, without heavy dressings, tomatoes, or too many other ingredients that tend to overpower the delicate wonder of good watermelon, my first favorite food. For a slightly more elaborate version, consult the Variation at the end of this recipe.

2 to 3 cups fresh ripe watermelon, in uneven cubes, seeded, chilled
3 handfuls small arugula leaves
Flake salt
Extra virgin olive oil
3 to 4 ounces feta cheese, broken into small pieces
Black pepper in a mill
⅓ cup Marcona almonds

Divide the watermelon among individual plates.

Put the arugula into a bowl, season lightly with salt, drizzle with a little olive oil and toss to coat thoroughly. Set on the plates next to the watermelon.

Scatter feta cheese over the watermelon, season with salt and several generous turns of black pepper and serve immediately.

Variation:

Toss the arugula with about ¼ cup very thinly sliced red onion, a handful of mint leaves and a handful of cilantro leaves. After tossing with salt and olive oil, add a generous squeeze of lime and several turns of black pepper. Divide among individual plates and scatter the watermelon, feta, and almonds on top of the greens. Season lightly with salt and generously with black pepper immediately before serving.

Chickpea, Celery, Green Olive, Scallion & Feta Salad

Serves 4 to 6

If you want this salad but don't have time to cook dried chickpeas, canned ones are a good substitute if you rinse them thoroughly before making the salad.

8 ounces dried chickpeas (garbanzo beans), soaked in water for several hours
Flake salt
4 ounces ditalini, pennette, or other small salad-style pasta
6 ounces feta cheese, crumbled coarsely
Grated zest of 2 limes
Juice of 2 limes
3 celery stalks, trimmed and cut into small dice
6 scallions, white and green parts, trimmed and very thinly sliced
½ cup pitted green olives, preferably Picholine, chopped
⅓ cup extra virgin olive oil, plus more as needed
2 tablespoons snipped chives
1 tablespoon minced Italian parsley
Black pepper in a mill

Drain the chickpeas, put them into a heavy saucepan, cover with water by 2 inches, add a tablespoon or so of salt and bring to a boil over high heat. Reduce the heat to medium-low and simmer gently until the chickpeas are completely tender, about 45 minutes. (Alternately, cook the chickpeas in a pressure cooker, according to the manufacturer's instructions.) Drain, rinse in cool water and drain thoroughly. Transfer to a wide shallow serving bowl and let cool to room temperature.

Cook the pasta in boiling salted water according to package direction until it is just tender. Drain, rinse in cool water, drain thoroughly and toss with the chickpeas. Add the feta, lime zest, and half of the lime juice and toss gently.

Add the celery, scallions, green olives, and olive oil and toss again. Taste for acid balance and add the remaining lime juice or a splash more olive oil as needed. Add the chives and parsley.

Add several very generous turns of black pepper, taste and correct for salt. Serve at room temperature, add several turns of black pepper, taste and correct for salt.

> **Variation:**
>
> For a main course salad, add canned and drained tuna, roasted chicken, or smoked trout after adding the celery, scallions, olives, and olive oil.

Carrot Salad with Black Olives, Pecans, Pomegranates & Feta

Serves 6 to 8

If you can find Nantes carrots or pale yellow carrots, use them, as they have a delicious and delicate sweetness that other carrots lack. When combined with salty olives and feta cheese and spiked with black pepper, they make a delicious salad. When pomegranates are in season, be sure to add some, as they dazzle both the eye and the palate.

2 pounds medium carrots, trimmed, peeled, and cut in ⅛-inch diagonal slices

4 ounces shelled pecans

2 to 3 ounces oil-cured black olives

Flake salt

1 tablespoon best-quality red wine vinegar

2 teaspoons ground cumin

Pinch of ground clove

4 tablespoons extra virgin olive oil, plus more to taste

2 tablespoons chopped fresh Italian parsley

4 ounces feta, broken into small pieces

⅓ cup fresh pomegranate arils

Black pepper in a mill

Put the carrots into a steamer basket set over boiling water, cover and steam until tender when pierced with a fork, about 7 to 9 minutes. Set aside to cool to room temperature.

Meanwhile, put the pecans into a small heavy sautè pan set over medium heat and toast until fragrant; do not let the pecans burn. Set aside.

Remove and discard the olive pits and chop the olives.

When the carrots have cooled, put them into a medium bowl and season lightly with salt. Add the vinegar, cumin and clove, toss, pour in the olive oil and toss again. Add the toasted pecans and the olives, along with the parsley, feta, and pomegranates.

Toss gently, season with several turns of black pepper, taste and correct for salt.

Transfer to a pretty bowl and serve at room temperature.

Poached Egg Salad with Frisee & Bacon Vinaigrette

Serves 2 to 4

Salads with poached eggs and bacon are the offspring of the classic French bistro dish, salade au lardon, *and there are countless versions. When the salad is seasoned perfectly with salt and pepper, when the eggs are poached so that the whites are tender but cooked through and the yolk hot but liquid, and when the vinegar is perfectly balanced with hot bacon fat and olive oil, it is one of the most delicious dishes in the world, as welcome at breakfast and dinner as it is at lunch.*

2 teaspoons white wine vinegar
3 to 4 handfuls frisee, rinsed and dried
Flake salt
4 large fresh farm eggs
Bacon Vinaigrette (page 360), hot
4 slices toasted and buttered hearth bread, optional

Fill a medium pan with about 3 to 4 inches of water, add the vinegar and bring to a boil over high heat.

While waiting for the water to boil, put the frisee in a bowl, season with salt, toss and divide among individual bowls or plates. Set aside. When the water boils, reduce it to medium heat so that it simmers gently.

Break an egg into a small bowl and slip it into the work. Working quickly, follow with the remaining eggs, keeping them separate. Set the timer for two minutes and adjust the heat so that the water simmers continuously but does not boil over. When the time sounds, turn off the heat and let rest 30 seconds to 1 minute (for very large eggs).

Drain off as much water as you can without disturbing the eggs and then use a slotted spoon to remove each one from the pan and set it on top of the frisee.

Spoon hot dressing on top and serve immediately, with or without toast alongside.

Smashed Deviled Egg Salad with Chermoula and Spring Greens

Serves 4 to 6

As very fresh farm eggs become more readily available, deviled eggs become a bit more difficult to make, as it can be a challenge to shell very fresh cooked eggs without breaking them apart. Faced with serving deviled eggs at my local farmers market on Easter morning, I quickly concocted an easier version when the first dozen eggs proved impossible to peel. I simply put the peeled but broken eggs into a large bowl and smashed them with a metal pastry cutter. At the market, I tossed fresh spring greens with a little grapefruit vinaigrette, topped the greens with the smashed egg salad and added a dollop of fiery Chermoula on top. It was a huge hit and I have since made this dish for dozens of people at farmers markets throughout the North San Francisco Bay where I live.

Preserved Lemon Chermoula,
   page 350
6 fresh local pastured eggs of equal
   size
1 small shallot, minced
Flake salt
1 tablespoon Champagne vinegar
2 tablespoons fresh grapefruit juice
Generous pinch of ground
   cardamom
Black pepper in a mill
3 to 4 tablespoons extra virgin

olive oil
3 tablespoons homemade or Best
   Foods mayonnaise
1 tablespoon Dijon mustard, plus
   more to taste
2 or 3 generous handfuls salad greens,
   preferably from Earthworker
   Farm
Kosher salt
Sourdough hearth bread or other
   rustic bread of choice, hot

First, make the Chermoula and set it aside.

To cook the eggs, put them in a saucepan, cover with water by at least 1 inch and set over medium-high heat. When the water boils, cover the pan and remove from the heat. Let rest from 7 to about 17 minutes, depending on the size of the eggs; the smallest ones need barely 7 minutes and the jumbos may need 17 or even 18 if they are cold when you begin.

Transfer the eggs to a ice water bath, lightly crack the shells on a hard surface and let sit in the cold water until cool.

While the eggs cook, make the dressing. Put the shallot into a small bowl, sprinkle with a little salt and add the vinegar, grapefruit juice, cardamom, and several turns of black pepper. Set aside for a few minutes. Stir in the olive oil, taste and correct for salt and acid. Set aside.

Crack open each egg and use a spoon to scoop out it out of its shell; if the eggs are very fresh, some of the white will likely stick to the shell; use a spoon to extract it.

Put the cracked eggs into a bowl, smash them with a fork or pastry cutter, add the mayonnaise and mustard and mix; the salad should be chunky.

Season to taste with salt and pepper.

To serve, put the salad greens into a bowl, add about half the dressing and toss. Save the remaining dressing to use another time.

Divide the greens among individual salad plates and top the greens with some of the egg salad. Add a dollop of Chermoula on top of the eggs and serve immediately, with the hot bread alongside.

Cobb Salad, Partially Deconstructed

Serves 4

Cobb Salad is typically served with rows of ingredients piled up against each other so that it quickly turns into a mess when it is served. It's also typically made with cubes of chicken breast which are dry and not very flavorful. I've always loved the idea of the salad more than the real thing and so I set out first to use thigh instead of breast meat and to deconstruct it a bit, so that it looks more like an abstract painting from the start, on a plate that is not too crowded. This version is relaxed, delicious, and pretty enough for a summer luncheon or dinner party.

Green Peppercorn Vinaigrette (page 155)
1 teaspoon Green Peppercorn Mustard or Dijon mustard, plus more to taste
Cooked meat from 2 chicken thigh-leg pieces, torn into bite-sized pieces
8 inner leaves butter lettuce, rinsed and patted dry
Flake salt
4 farm eggs, hard cooked and peeled
4 bacon slices, cooked until crisp, drained and coarsely chopped
2 ounces blue cheese, such as Point Reyes Original Blue, broken into bite-
 sized chunks
1 firm-ripe Haas avocado, cut in half, each half cut into thin cross-wise slices
3 cups fresh watercress, stems trimmed, chopped
1 ¼ cups quartered cherry tomatoes, preferably a variety of colors
Black pepper in a mill

First, make the dressing, add the mustard and set aside.

Put the chicken into a small bowl, spoon a generous tablespoon of dressing over it, toss and set aside.

Set 4 large plates near your work surface and set 2 butter lettuce leaves off center on each plate. Season with a little salt. Mound the chicken on the leaves, so that it tumbles off a bit.

Set the eggs on your work surface, crush each one once with a dinner fork, lift it carefully and set near the center of the plate; scatter some bacon nearby but not too close.

Add some blue cheese a fair distance from the other ingredients and place some avocado on the plate, not too close to anything else, and press to fan out the slices.

Hold watercress above the center of the plate and let it fall randomly. Repeat with the tomatoes.

Season all over with black pepper and season everything but the cheese and bacon with salt.

Spoon dressing over the salad and serve right away.

Green Peppercorn Vinaigrette

2 tablespoons white wine vinegar
1 tablespoon minced chives or 1 small shallot, minced
2 teaspoons crushed dried green peppercorns
1/2 teaspoon crushed black peppercorns
1 to 2 teaspoons kosher salt
1 egg yolk, optional
1/3 to 1/2 cup extra virgin olive oil

In a small bowl, combine the vinegar, chives, peppercorns, and salt. Let the mixture sit for 15 minutes. Whisk in the egg yolk, if using; slowly whisk in the olive oil. Taste and correct the seasoning. Use immediately, or refrigerate, covered, until ready to use. Use within 3 to 4 days.

Dungeness Crab Salad

Serves 2, easily doubled

Dungeness crab is king on the West Coast, where its region stretches from Point Conception in Santa Barbara County, California, all the way to Alaska. Other species of crab will work beautifully in this salad and I always recommend using what is closest to where you live. You can also use the small squid typically called calamari, small clams, mussels, or a combination of all three.

Chilled meat from 1 large Dungeness crab
1 small shallot, minced
½ serrano, seeded and minced
1 teaspoon brined green peppercorns, drained and cracked
Pinch of chipotle powder or smoked paprika
Flake salt
Black pepper in a mill
2 tablespoons freshly squeezed lemon juice
1 celery stalk, trimmed and cut into very small dice
3 to 4 tablespoons extra virgin olive oil
2 tablespoons fresh cilantro or Italian parsley leaves, chopped
1 small head butter lettuce, inner leaves only
2 lemon wedges, for garnish

First, prepare the crab if you have not already done so. Chill thoroughly.

Put the shallot, jalapeno or serrano, green peppercorns, chipotle powder or smoked paprika, a generous pinch of salt, and several generous turns of black pepper into a small bowl. Add the lemon juice, agitate the bowl and set aside for 20 to 30 minutes.

To finish the salad, put the crab into a medium bowl, add the celery and 3 tablespoons of the olive oil, and toss very gently. Pour the lemon juice mixture in, add the cilantro or parsley, and toss again.

Taste and correct for acid, adding the remaining tablespoon of olive oil if needed for balance. Taste and correct for salt.

Divide the lettuce between two plates, spoon the salad on top, garnish with a lemon wedge, season with a few more turns of pepper and serve immediately.

A Salt & Pepper Cookbook

Salt Cod Vinaigrette

Serves 4 to 6

Salt cod is an essential part of traditional Portuguese cooking, and it is easy to find it in eastern New England, where there is a large Portuguese population, primarily from the Azores. You'll find it on the menu of the region's restaurants, too, such as Mike's Kitchen in Cranston, known as The Post to its loyal customers, primarily members of the Veterans of Foreign War, which shares quarters with the restaurant. In the late 1990s, The Post was discovered by students at nearby Brown University and the Rhode Island School of Design, who enjoy the unselfconscious retro atmosphere and low prices. Portions are huge, waitresses seem to have stepped out of a 1950s truck stop, and the quality of everything, from salt cod vinaigrette to creamy polenta with sausages, is sublime.

1 pound salt cod, soaked for
   1 or 2 days (see page 112)
¼ cup white wine vinegar
1 tablespoon lemon juice
1 teaspoon flake salt
1 tablespoon minced fresh flat-leaf
   parsley

1 teaspoon fresh oregano leaves
1 teaspoon fresh thyme leaves
⅔ cup olive oil
Black pepper in a mill
1 head butter lettuce, washed and
   leaves separated
1 lemon, cut into wedges

Remove the salt cod from its final cold-water bath, place it in a large sauté pan, cover with fresh cold water, and set over medium heat. When the water boils, remove the pan from the fire, and cover it. Let the fish sit for 15 minutes, then drain it thoroughly and allow it to cool. Break the fish into small pieces and pick out any pieces of bone and set aside.

To make the vinaigrette, whisk together the vinegar, lemon juice, and salt. Add the parsley, oregano, and thyme, and whisk in the olive oil. Season with black pepper.

Arrange the lettuce on individual serving plates. Toss the salt cod with the dressing and spoon some of it on top of each serving of lettuce. Garnish

with lemon wedges and serve immediately.

Shrimp, Avocado & Arugula Salad with Radish Vinaigrette

Serves 3 to 4

It is only recently that radishes have been sold with beautiful pert greens still attached. For decades, most radishes were either sold separated from their greens or attached to greens that were so wilted you barely wanted to touch them, let alone eat them. But with the growing popularity of farmers markets, just-harvested radishes with the peppery greens in perfect condition are readily available nearly year round. I use them to make delicious soups and add them to other greens for a quick sautè. Here, I add them to arugula to add yet another peppery flourish to this simple, refreshing, and filling salad.

1 small shallot, minced
2 to 3 radishes, trimmed and minced
1 tablespoon snipped fresh chives
Flake salt
3 tablespoons freshly squeezed lime juice
Black pepper in a mill
5 to 6 tablespoons best-quality extra virgin olive oil
Radish leaves, if pristine
3 generous handfuls small-leafed arugula
1 firm-ripe avocado, pitted, peeled, and cut into thin lengthwise slices
8 ounces baby baby shrimp, preferably from Oregon

Put the shallot, radishes, and chives into a small bowl. Season with salt, add the lime and agitate the bowl a bit to dissolve the salt. Add a few turns of black pepper and whisk in the olive oil. Taste and correct for salt and acid. Set aside briefly.

Put the radish greens, if using, and arugula into a medium mixing bowl, season lightly with salt and add about a tablespoon of dressing. Toss and divide among individual plates.

Add avocado, fanning the slices off center. Scatter the shrimp on top, making sure some are on the avocado.

Spoon the dressing over the salad and serve immediately.

Salt Cod, Potato & Artichoke Salad

Serves 4 to 8

Once you've learned your way around salt cod, why limit yourself to a single recipe or two? Here I made a more elaborate version of Salt Cod Vinaigrette, with earthy artichokes, peppery radishes, and salty capers and olives.

¾ pound salt cod, soaked in cold water for 1 or 2 days (see page 112)
4 large or 6 medium artichokes, trimmed
4 teaspoons kosher salt
1 tablespoon olive oil
8 Yukon Gold potatoes (or other waxy-fleshed variety), each about 3 inches long, peeled
¼ cup sherry vinegar
2 to 3 tablespoons lemon juice
3 garlic cloves, minced
Black pepper in a mill
2 tablespoons minced fresh flat-leaf parsley
1 teaspoon fresh tarragon leaves, minced
1 teaspoon fresh chervil leaves, minced
⅔ cup extra virgin olive oil
5 large eggs, hard-cooked and peeled
1 bunch (about 10) radishes, trimmed and thinly sliced
¾ cup cracked green olives, pitted and sliced lengthwise
2 tablespoons capers, drained

Remove the salt cod from its final cold-water bath, place it in a large sauté pan, cover with fresh cold water, and set over medium heat. When the water boils, remove the pan from the fire, and cover it. Let the fish sit for 15 minutes, then drain it thoroughly and allow it to cool. Break the fish into small pieces and pick out any pieces of bone. Set aside or cover and refrigerate.

Place the artichokes in a large pot of water, add 3 teaspoons of the kosher salt, and drizzle a little of the olive oil on top of each artichoke. Bring to a

boil, cover, and cook until tender, from 20 to 40 minutes depending on the size and type of artichoke. Using tongs or a slotted spoon, transfer the artichokes to a strainer or colander to cool. Keep the cooking liquid in the pot.

Cut the potatoes in half lengthwise and then cut each half into slices about ⅜ inch wide; do not make them too narrow. Put the potatoes into the artichoke cooking water, bring to a boil, reduce to a simmer, and cook until the potatoes are tender, about 20 minutes. Drain thoroughly and set aside to cool.

Meanwhile, make the dressing. In a small bowl, whisk together the vinegar, lemon juice, garlic, the remaining teaspoon of kosher salt, and several turns of black pepper. Add the parsley, tarragon, and chervil, and whisk in the olive oil. Taste the dressing, correct the seasoning, and set aside.

When the artichokes are cool enough to handle, separate the leaves from the hearts and scoop out and discard the thistle-like choke in the center. Cut the meat from the base of each artichoke into thin julienne (on the outer leaves you will get just one strip; the inner leaves will be tender enough for you to make 3 or 4 strips). Cut the hearts crosswise into ¼-inch slices.

Cut 4 of the eggs into quarters. To assemble the salad, place the potatoes, artichokes, radishes, salt cod, sliced eggs, and olives in a large bowl.

Add half the dressing and toss together gently. Transfer the salad to a serving platter and spoon the remaining dressing over it. Using an egg slicer, cut the remaining egg into round slices and set them on top of the salad. Scatter the capers and grind some pepper over the salad, and serve immediately.

Pasta, Rice, Polenta & Other Grains

Salt is an essential ingredient in coaxing out the flavor of nearly all grains. When making risotto, for example, the onions and other aromatics should be salted before rice is added, and the final dish must be salted, too, or it won't come together. If steamed rice will be served neat, without condiments, the cooking water must be salted.

Most breads include salt, which both controls the action of the yeast and adds flavor. The breads of Tuscany are exceptions; Tuscan bread is by tradition not salted, but salt is added at the table, as a condiment. Pepper is used to flavor flatbreads in countries where pepper is grown, including India, where the crispy lentil wafers called pappadums are sometimes spiked with ground black pepper. *Botia roti* is a fresh flatbread, seasoned with black pepper, salt, and other spices.

Salted Mustard Greens & Rice

Spaghetti Carbonara

Spaghetti with Pepato, Black Pepper & Nutmeg

Spring Linguine with Fresh Artichokes, Favas & Feta

Dried Gnocchi with Ricotta, Peppercorns, Lemon & Basil

Black Pasta with Butter & Golden Caviar

Risotto Pepato

Risotto with Zucchini, Green Peppercorns & Basil

Bram's Egyptian Baked Rice with Black Pepper

Malaysian Chicken Rice

Summer Farro & Bean Salad with Avocado, Tomato, Feta & Sorrel

Red Beans & Rice

Black Pepper Polenta Made in a Slow Cooker

Three-Peppercorn Bread with Pancetta & Garlic

Pepper-Crusted Pizza with Porcini, Fontina & Sage

Savory Zucchini Galette

Savory Galette Dough

Cooking Perfect Pasta

Perhaps nowhere do we see the essential nature of salt function more clearly than in the cooking of pasta. The cooking water must be properly salted. You cannot add flavor that fails to develop because of a lack of salt. I add 4 tablespoons of kosher salt to 6 quarts of water and find that with less than this amount, the pasta tastes bland. In some cases (if, for example, your water has a lot of sodium in it), 3 tablespoons will be sufficient.

You must begin with sufficient water if you want your pasta perfectly cooked; it needs room to tumble freely as it absorbs water and expands. If you have one of those handy pasta pots, you won't have questions about size, but if you don't, use the largest pot you have. The best is a heavy stockpot that holds 6 quarts of water with 3 inches of space above the water. This is large enough to cook up to 2 pounds of strand (noodle) pasta or large shapes, or 3 pounds of small shapes, such as farfalline.

Allow the water to come to a rolling boil, and add the salt after the water boils; adding it first really doesn't shorten the time it takes to boil the water. Some cooks believe that adding the salt before the water boils causes an aftertaste in the pasta, but I have not found that to be the case and there is no explanation for such a possibility. I add salt just before I add the pasta so that I don't have to remember if I've already added it.

Stir the pasta after putting it in the water, and stir now and then as it cooks to be certain it is not sticking together. Pay attention to the cooking time recommended on the package, but begin tasting it a minute or two before you expect it to be done.

Drain the pasta in a large strainer or colander the instant it loses its raw firmness yet retains a bit of resistance, that is to say, when it is al dente. If you will be serving the pasta immediately, do not rinse it. If you are preparing the pasta for a recipe that will undergo further cooking, or for a dish that will be served at room temperature, rinse it in cool water and toss it with a small amount of olive oil to prevent it from sticking. Place pasta to be held until later in a wide bowl so that it is not be crushed by its own weight.

Salted Mustard Greens & Rice

Serves 4 to 6

Salting is a technique used to preserve a variety of greens all over the world; fermented (such as sauerkraut) or not (such as these), mustard greens have a tangy spiciness and a crispness that is extremely appealing. Here, they are tossed with aromatic rice for a simple yet satisfying vegetarian dish.

2 cups jasmine rice
Kosher salt
3 tablespoons toasted sesame oil
1 ½ cups, packed, Salted Mustard
    Greens (page 279), coarsely
    chopped

Kosher salt
Black pepper in a mill
1 cup (from 1 large or 2 small)
    chopped ripe tomato
1 teaspoon Gomashio (page 348)

Put the rice in a strainer or colander and rinse it under cool running water until the water runs clear. Transfer it to a large saucepan, add 3 cups cold water and a generous pinch of salt and bring to a boil over high heat. Reduce the heat to low, cover the pan, and cook for 20 minutes without lifting the lid. Remove from the heat, let the rice steam undisturbed for 10 minutes, and fluff it with a fork.

Heat the sesame oil in a large sauté pan set over medium heat, add the cooked rice and stir to heat through. Add the greens and continue to cook, tossing and stirring, until the greens are hot.

Season to taste with salt and pepper. Divide among individual serving bowls, top each portion with some of the tomatoes and a little of the gomashio, and serve immediately.

Variation:

When tomatoes are not in season, serve with Spicy Salt Egg Sauce (page 355).

Spaghetti Carbonara

Serves 3 to 4 as a first course, 2 as a main course

No book about salt and pepper would be complete without a recipe for the classic Roman dish, Spaghetti Carbonara—coal miner's spaghetti—named for the black pepper that punctuates the strands of pasta like bits of black coal. You'll find many versions of this dish, some with cream, others with mushrooms, onions, prosciutto, and olives, a few with barely any pepper at all. I've never had a bad version, though I admit disappointment when the dish arrives with cream, as it eclipses the rich, silky texture of the classic version. Then it's a good dish, but more closely resembles Fettuccine Alfredo than Spaghetti Carbonara.

When I worked as a restaurant critic, I once took an Italian café to task for serving just such a version, full of cream, while describing it as classic carbonara. After the review appeared, I received a delightfully defensive letter from the restaurateur who blamed the cream on the presence of American troops in Italy during World War II, who, he claimed, demanded cream in everything. "Cream is communism!" he declared, the great equalizer that renders distinct flavors similar. Everyone demands it these days, he continued, adding a promise to prepare the classic version on my next visit.

Mostly, I enjoy it at home, as I love my version.

8 to 10 ounces best-quality dry bucatini
Kosher salt
1 tablespoon olive oil
3 garlic cloves, peeled
2 to 3 ounces pancetta, cut into small dice
2 large farm eggs (3 if small)
2 ounces Pepato or other peppered cheese, grated
2 ounces Parmigiano-Reggiano, grated
Best-quality black pepper in a mill
Sel gris or fleur de sel

Bring a large pot of water to a boil, add a generous 3 tablespoons of salt, stir in the bucatini and cook according to package directions until just done. Stir now and then as it cooks. Drain but do not rinse the pasta.

Meanwhile, pour the olive oil into a small skillet, add the garlic, and sauté for about 90 seconds; do not let it brown. Add the pancetta and sauté until it loses its raw look but is not fully crisp. Remove from the heat and let cool slightly.

Break the eggs into a deep bowl and whisk until very smooth and creamy. Add the cheese and whisk again. Tip in the garlic and pancetta and add a lot of black pepper, at least a full tablespoon (though it is not necessary to measure it).

When the pasta is done, *immediately* tip it into the bowl with the egg mixture and use two forks to lift it over and over until each strand is fully coated with the mixture.

Divide among individual pasta bowls, add several turns of black pepper and a sprinkling of salt to each portion and serve immediately.

Spaghetti with Pepato, Black Pepper & Nutmeg

Serves 3 to 4

I enjoy this dish around the winter holidays, when hectic schedules and rich foods threaten to eclipse our normal lives. Then, I like nothing better than a simple Caesar salad followed by this lean and fragrant pasta. Add a tangy cranberry soup and bottle of Champagne and you have an elegant Christmas Eve dinner that is very simple to prepare. It is important to use freshly grated nutmeg in this dish and there is now an easy way to do so. For years, I skinned my knuckles nearly every time I used one of those tiny aluminum nutmeg graters. Now Zassenhaus, producer of outstanding pepper grinders, has developed a hand-cranked nutmeg mill. It looks nearly exactly like a pepper mill, but with a compartment on the bottom that holds a single whole nutmeg. It is a marvelous invention; you'll never need to buy ground nutmeg or bleed into your pasta again.

12 ounces spaghetti or other thin strand pasta
3 ounces pepato (see Note on next page)
6 tablespoons extra virgin olive oil
2 teaspoons kosher salt
Black pepper in a mill
Nutmeg in a grinder

Cook the pasta in boiling salted water until it is just done. While the pasta cooks, prepare the cheese. Use a vegetable peeler to make several curls of cheese, set them aside, and grate the remaining cheese using the fine blade of a grater.

Drain the pasta thoroughly and place it in a medium bowl. Drizzle the olive oil over the pasta and toss thoroughly; add the grated cheese and toss again. Sprinkle with salt, grind generous amounts of both pepper and nutmeg over the pasta, and toss lightly and quickly. Divide among warm plates, top each portion with a few curls of cheese, and serve immediately.

NOTE

Pepato, sometimes called Pecorino Pepato, is a sheep's milk cheese studded with peppercorns. It is traditional in Italy but is now made in the United States, including by Bellwether Farms, located in Sonoma County, California.

Variation:

Divide 4 cups of mixed young salad greens (such as arugula, cress, nasturtium leaves, miner's lettuce) among 3 or 4 individual plates. After tossing the pasta with the olive oil, cheese, salt, pepper, and nutmeg, set some on top of each portion of greens. Garnish with the curls of cheese and serve immediately.

Spring Linguine with Fresh Artichokes, Favas & Feta

Serves 4 to 6

Spring favas, sweet and delicate, add bright notes to this springtime dish, artichokes add an earthy savor, and salty feta cheese, punctuated with aromatic black peppers, brings everything together. You can make this dish in the fall, as well, but you'll need to do without the favas, as they are a singularly spring vegetable.

4 medium artichokes, cooked (see Note on next page)
1 cup fresh fava beans, shelled, blanched, and peeled
1 pound fresh linguine or 12 ounces dried linguine
6 ounces French, Bulgarian, or Greek feta, cut into ¼-inch cubes
4 tablespoons extra virgin olive oil, plus more to pass at table
2 teaspoons black peppercorns, coarsely crushed
Maldon Salt Flakes
2 tablespoons snipped chives or minced fresh flat-leaf parsley

Cook the artichokes if you have not already done so. After the artichokes are cool to the touch, separate the leaves from the hearts, cut out the thistle-like choke, and discard it. Cut the hearts crosswise into thin slices, and then cut the slices in half. Using a very sharp knife, cut the meat at the base of each artichoke leaf into thin julienne. Put the sliced hearts and julienned leaf tips in a large bowl and set aside.

Prepare the fava beans, as well, and add them to the bowl with the artichokes.

Bring a large pot of water to a boil. When it reaches a rolling boil, add 3 generous tablespoons of kosher salt, stir in the pasta and cook according to package directions until just done. Drain but do not rinse.

While the pasta cooks, add the feta, olive oil, crushed peppercorns, and lemon zest to the artichokes, and toss with a fork. Add the drained pasta to to the bowl, and use two forks to toss it thoroughly but gently.

Taste and correct for salt. Divide among individual pasta bowls, season with a little Maldon salt flakes, scatter chives on top and serve immediately, with extra virgin olive oil alongside for drizzling.

> ## NOTE
>
> To cook artichokes, fill a large pot two-thirds full with water and add a tablespoon of salt. Pour a teaspoon of olive oil into the top of each artichoke, so that it sinks down into it. When the water boils, use tongs to place the artichokes in the water. Return to a boil, reduce the heat, and simmer, partially covered, until you can easily pull off a leaf, about 15 to 35 minutes, depending on the age and size of the artichoke. Drain the artichokes and them let cool to room temperature.

Dried Gnocchi with Ricotta Peppercorns, Lemon & Basil

Serves 4 to 6

Two peppers—good black pepper, freshly ground so that it is as fragrant as possible, and tangy green peppercorns in brine—offer a sensual contrast to the refreshing tastes of lemon, basil, and ricotta in this remarkably simple recipe. Ricotta is made from the whey that remains after a firm cheese is made; the whey is cooked with vinegar, which causes soft curds to form and rise to the top. The curds are scooped out, salted or not—the best, such as that made in Sicily, is salted—and packed in containers with holes in them that allow the cheese to drain. Ricotta is best eaten shortly after it has been made, but it's nearly impossible to find it this fresh outside Italy. It's another thing entirely from the cloying, dense packaged stuff that most of us think of as ricotta. Fresh-style ricotta is available in the United States, most commonly in areas with large Italian communities; do not use standard commercial or low-fat ricotta.

1 pound dried gnocchi (or other medium-sized pasta shape, not strand)
9 ounces fresh ricotta
Grated zest from 2 or 3 lemons
4 garlic cloves, minced
1 tablespoon green peppercorns in brine, drained
2 teaspoons freshly crushed black pepper
1 teaspoon kosher salt, plus more to taste
¼ cup fresh basil leaves (loosely packed), torn into small pieces
2 tablespoons snipped chives
4 tablespoons Meyer lemon olive oil (see Note on next page)
4 tablespoons homemade bread crumbs, lightly toasted

Bring a large pot of salted water to a boil and cook the pasta according to package directions. Meanwhile, place the ricotta in a large bowl, add the lemon zest, garlic, green peppercorns, black pepper, and the teaspoon of salt.

Drain the pasta quickly but not thoroughly; you want some of the cooking water to cling to it. Add the pasta to the ricotta, toss thoroughly so that the cheese melts from the heat of the pasta, add the basil, chives, and olive oil, and toss again. Season lightly with salt, scatter the bread crumbs on top, and serve immediately.

NOTE

O Olive Oil was the first Meyer lemon olive oil, made in northern California using ripe Mission olives and organic Meyer lemons. In the United States there are now several good brands, including DaVero, McEvoy, and The Olive Press, all located in the Northern California. A good substitute is 3 tablespoons extra virgin olive oil mixed with 1 tablespoon fresh lemon juice and 1 teaspoon minced lemon zest.

Black Pasta with Butter & Golden Caviar

Serves 4

All salt is, in the end, sea salt, the oldest the essence of a long-ago saltwater ocean, the newest freshly evaporated seawater. Caviar, too, evokes the sea, as does squid ink, which is responsible for the deeply colored pasta in this recipe. Until recently, it was almost impossible to find dried black pasta, but now several Italian companies make it and it is readily available in most parts of the United States. Fresh black pasta retains more of an aroma of the sea, so check with local pasta shops—it is often made around Halloween. Lima sea salt has a purplish gray sheen to it, which is why I recommend it here; feel free to use kosher salt or a small (not fine or ground) sea salt crystal in its place.

12 ounces black pasta
4 tablespoons unsalted butter, at room temperature
Unrefined sea salt, such as Lima
Black pepper in a mill
6 ounces American golden caviar
3 tablespoons minced red onion
2 tablespoons minced fresh flat-leaf parsley
1 hard-cooked egg, sieved
1 lemon, cut into wedges

Cook pasta until just done in plenty of rapidly boiling salted water. Drain the pasta but do not rinse, and place it in a large bowl.

Working quickly, add the butter, season with salt and pepper, and toss quickly until the butter is melted. Add half the caviar, toss again, and divide among soup bowls or plates. Top each portion with some of the remaining caviar, red onion, parsley, and egg. Garnish with lemon wedges and serve immediately.

Risotto Pepato

Serves 4

Three varieties of Italian rice, Arborio, Carnaroli and Vialone Nano, are used to make traditional risotto. Arborio is the most common and easiest to find but it is worth the effort to look for the others, especially Vialone Nano, which is more delicate and creamier. If you have an Italian deli near you, you should be able to find it. Think of this recipe as a savory canvas for the flavor of peppercorns, in both the pepato, and added to the risotto itself. Use the best quality peppercorns you can find; crushing them shortly before using them is essential to retain their full aroma and flavor.

2 tablespoons unsalted butter
2 medium leeks, white and pale, green parts only, trimmed, thoroughly cleaned, and cut into thin rounds
Kosher salt
1 shallot, minced
3 garlic cloves, minced

1 ¼ cups Vialone Nano or Carnaroli rice
6 to 7 cups chicken stock, chicken broth, or vegetable broth, hot
4 ounces pepato, grated
2 teaspoons freshly cracked black peppercorns
2 ounces pepato, in one piece

Heat the butter in a large sauté pan over medium heat until it is completely melted. Add the leeks and sauté until they are completely wilted, about 10 minutes. Season with about ¾ teaspoon salt, add the shallot and garlic, and cook for 5 minutes more. Add the rice and stir with a wooden spoon until each grain begins to turn milky white, about 3 minutes.

Keep the stock warm in a pot over low heat. Add the stock half a cup at a time, stirring after each addition until the liquid is nearly absorbed.

Continue to add stock and stir until the rice is tender, about 18 minutes. Just before the last addition of stock, stir in the grated pepato, add the pepper, taste, and correct the seasoning. Stir in the last of the liquid and ladle the risotto into warm soup plates. Using a vegetable peeler, quickly make curls of the remaining pepato, Scatter them over the risotto and serve immediately.

A Salt & Pepper Cookbook

Risotto with Zucchini, Green Peppercorns & Basil

Serves 3 to 4

In this risotto, dried green peppercorns add a bright, fresh taste and brined green peppercorns add a tangy spark, flavors that resonate beautifully with fresh zucchini and summer basil. This dish is best in early summer, when the first little zucchinis appear, their flavors so delicate and welcome after a winter without them. Basil, too, is at its best in early summer.

4 tablespoons clarified butter
¾ pound zucchini, cut in ¼-inch dice
1 tablespoon crushed dried green peppercorns
2 teaspoons freshly crushed black peppercorns
Kosher salt
1 small (about 4 inches) zucchini, cut into very thin julienne
2 tablespoons olive oil
1 small yellow onion, diced
1 teaspoon minced fresh garlic
1 ¼ cups Vialone Nano or Carnaroli rice
6 to 7 cups chicken stock, chicken broth, or vegetable broth, hot
1 tablespoon fresh lemon juice
3 ounces grated Dry Jack Cheese
1 tablespoon brined green peppercorns, drained
3 tablespoons minced fresh basil leaves

Heat 2 tablespoons of the butter in a medium skillet over medium heat. Add the diced zucchini and sauté it until it is just tender, about 8 minutes. Add 1 teaspoon each of the peppercorns and season with about ¾ teaspoon salt. Transfer the zucchini to a bowl and set aside. Melt a teaspoon of the remaining butter in the sauté pan, add the julienned zucchini and cook quickly until just barely limp, about 2 or 3 minutes. Season with small pinches of green peppercorns, black peppercorns, and salt and set aside.

Heat the remaining butter and the olive oil in a large sauté pan over medium heat. Add the onion and sauté until soft and fragrant, about 8 minutes. Add the garlic and sauté for 2 minutes more. Add the rice and stir with a wooden spoon until each grain begins to turn milky white, about 2 minutes. Keep the stock warm in a pot over low heat. Add the stock, half a cup at a time, stirring after each addition until the liquid is nearly absorbed.

Continue to add stock and stir until the rice is tender, about 18 minutes. Stir in the diced zucchini, lemon juice, remaining green and black peppercorns, and 2 tablespoons of the basil. Fold in the cheese and the brined peppercorns.

Taste, correct the seasoning, and remove from the heat.

Working quickly, reheat the julienned zucchini. Ladle the risotto into individual soup plates, top each portion with some of the julienned zucchini and some of the basil. Serve immediately.

Bram's Egyptian Baked Rice with Black Pepper

Serves 4 to 6

A bram is a deep open clay vessel from Egypt, where it is used as a to-go container in restaurants. It is also the name of a lovely little shop, Bram Cookware, located in the town of Sonoma in Northern California, which imports brams and a wide array of other clay pots, including beautiful hand-carved and hand-painted tagines. This traditional Egyptian dish is both easy to make and wildly delicious. Serve it with roasted chicken, curries, or alongside anything that calls for simple rice.

2 cups short grain or sushi rice
2 teaspoons sel gris or other solar-dried unrefined salt
4 cups whole milk, preferably organic
1 cup heavy cream
Black pepper in a mill
3 tablespoons butter, preferably organic, cold

Put the rice into the bram, add the salt and the milk and stir. Season very generously with black pepper.

Slice the butter into thin shards and scatter them on top of the rice mixture.

Set a baking sheet on the middle rack of the oven, set the bram on top and set the temperature to 375 degrees.

After 30 minutes, check the rice to be certain it is not bubbling over.

Cook until the milk is fully absorbed by the rice and there is a golden brown crust over it.

Remove from the oven and serve as a side dish.

Malaysian Chicken Rice
Nasi Ayam

Serves 3 to 4

During my first trip to Malaysia, Ramlee and Talib, the guides who were taking me upland to visit peppercorn farms, repeatedly asked if I wanted chicken rice for lunch. "What's the big deal?" I wondered, envisioning something along the lines of Chinese pork fried rice, but with chicken. Yet they seemed so eager that I said "sure."

In Sarikei, we pulled into a parking place in front of a restaurant called "Restoran Malaysia Chicken Rice." They served a single dish and it bore no resemblance whatsoever to what I had expected.

Our identical lunches came quickly, their evocative aromas announcing their arrival. Each plate was filled with white rice and plump chicken slathered in a dark sauce, with three little bowls of condiments and a slightly bigger bowl of broth alongside.

First, I sipped the chicken broth after adding a bit of each condiment, following my hosts' lead. Soon, I bit into a juicy chicken thigh and in that moment the world was transformed. I had never had such delicious chicken, with firm yet utterly tender muscle tone and true chicken flavor.

Soon I was noticing that chicken rice was everywhere, in brick and mortar restaurants, in hawker centers, in food courts, and on billboards. When I returned to Kuala Lumpur, I walked to one of the city's largest shopping malls, where the top floor was devoted to traditional Malaysian cuisine. Most tourists never made it above the floor that included MacDonald's, Burger King, and KFC and so I savored another succulent version of chicken rice with Malaysians, who seemed a tad amused by my obvious delight.

The secret to the chicken was, I believe, its absolute freshness, which was highlighted by the sweet and savory sauces that came with it.

This recipe is as close as I've been able to get to the flavors of those faraway meals. To make it absolutely authentic, begin your meal with a bowl of clear chicken broth spiked with a bit of ginger, lemongrass, and garlicky hot sauce.

for the marinade
6 garlic cloves
2 or 3 shallots
One 2-inch piece fresh ginger, peeled and grated
2 tablespoons black soy sauce
2 tablespoons light soy sauce
2 tablespoons oyster sauce
1 tablespoon bottled Thai chili sauce
2 teaspoons kosher salt
1 teaspoon chili powder
1 teaspoon ground white pepper

1 large (4 to 5 pounds) whole pastured chicken, as fresh as possible, cut into pieces

for the condiments
One 3-inch piece fresh ginger, peeled and grated
2 tablespoons mild olive oil (see Note)
¼ cup black soy sauce
1 tablespoon brown sugar
3 garlic cloves, minced
1 small shallot, minced
1 teaspoon crushed red pepper flakes
1 tablespoon fresh cilantro leaves

to serve
4 cups cooked long-grain white rice, hot (see Variation)
1 cucumber, peeled and sliced thin
Thai chili sauce in a small serving bowl

To make the marinade, use a mortar and pestle or a suribachi and pound the garlic and the shallots until both are reduced nearly to a pulp. Add the ginger and pound together. Transfer the mixture to a medium bowl, stir in the soy sauces, oyster sauce, chili sauce, salt, chili powder, and pepper. Set aside.

Rinse the chicken under cool water and pat it dry it with a tea towel. Prick the skin of the chicken with a fork or the tip of a very sharp knife. Using your hands, rub the marinade into the chicken, set the chicken in a single layer in a glass baking dish, cover, and refrigerate for at least 4 hours or overnight.

To finish, preheat the oven to 450 degrees.

Place the chicken pieces on a roasting rack set on a baking sheet and roast for 20 to 25 minutes, until the chicken is cooked through, to about 145 to 150 degrees (the temperature will rise as the chicken rests out of the oven).

Meanwhile, mix together the grated ginger and olive oil in a small bowl and set it aside. In a small saucepan, combine the soy sauce, brown sugar, garlic, shallot, and red pepper flakes. Cook over low heat, stirring constantly, until the sugar is dissolved. Transfer to a small bowl, add the cilantro leaves, and set aside.

Remove the chicken from the oven, cover lightly with parchment and let rest for 15 minutes. Use a cleaver to hack the chicken through the bone into pieces about 2-inches wide. (If you don't have a cleaver or don't feel comfortable using one, leave the chicken pieces whole.) Place the cooked rice on a large serving platter, add the chicken to the platter, and garnish around the edges with cucumber. Serve the chicken immediately, along with the grated ginger, seasoned soy sauce, and Thai chili sauce. Guests dip pieces of chicken into the condiments.

Variation:

Before cooking the rice, add a teaspoon of pounded fresh ginger, a teaspoon of pounded garlic, 1 tablespoon of butter, and 2 strips of pandan (screwpine) leaves to the cooking water. Top the rice with ¼ cup fried shallots.

NOTE

In Malaysia, as in much of Southeast Asia, coconut palm oil is the most commonly used fat. It has a very mild taste; you can substitute a mild olive oil if you like, but you want to avoid the additional flavor that would come from a fruity olive oil. Organic coconut palm oil, one of the world's healthier oils, is increasingly available in Asian markets and supermarkets.

A Salt & Pepper Cookbook

Summer Farro & Bean Salad with Avocado, Tomato, Feta & Sorrel

Serves 4 to 6

As we delve deeper, as a culture, into what we call heirloom foods—casually put, foods that have been overlooked and forgotten during decades of reliance on corporate agriculture—the array of foods available to us has expanded exponentially. Farro, once enjoyed only in Europe and then found here and there in Italian markets, is now readily available in most parts of the country. Domestic farmers are growing it, too, and not for the first time. Decades ago, our farmers grew emmer, as farro is called here. It is a wonderful grain, with a rich nutty flavor and chewy texture. I find it best when seasoned with salt, pepper, and lemon juice while still warm.

Use this recipe as a template, varying ingredients based on what is in season.

4 ounces dried beans, such as flageolet or cannellini, soaked in water
    overnight
Kosher salt
4 ounces semi-pearled farro, soaked in water overnight
Kosher salt
Juice of 2 to 3 lemons
Black pepper in a mill
1 firm-ripe avocado, peeled and cubed
1 medium tomato, preferably orange, chopped
1 small red onion, cut into small dice
½ cup extra virgin olive oil, plus more to taste
4 ounces feta cheese, crumbled
1 cup shredded fresh sorrel or small arugula
¼ cup chopped fresh Italian parsley

Drain the beans, cover with fresh water by 2 inches and bring to a boil over high heat. Skim off any foam that rises to the surface. Reduce the heat and simmer until tender, about 35 minutes. When the beans are nearly tender and beginning to give off a bean-like aroma, season them with salt. Drain the cooked beans and transfer to a wide bowl to cool.

Drain the farro, cover it with fresh water by 2 inches, add a generous teaspoon of salt, and bring to a boil over high heat. Skim off any foam that rises to the surface. Reduce the heat and simmer until tender, about 30 minutes. Drain the farro and transfer to a large wide bowl to cool. While still hot, drizzle with the juice of 1 lemon and several turns of black pepper and toss.

When the beans and farro have cooled to room temperature, add the beans to the farro, along with the avocado, tomato, and onion. Add half the remaining lemon juice and the ½ cup of olive oil and toss gently. Taste for acid balance, adding more lemon juice and more olive oil as needed.

Season with several generous turns of black pepper, add the feta cheese, sorrel or arugula, and parsley and toss again. Taste and correct for salt.

Serve at room temperature.

Red Beans & Rice

Serves 6 to 8

You may be surprised to see garlic powder, onion powder, dried spices, and, yes, commercially ground white and black pepper, ingredients increasingly eschewed as we embrace all things as fresh and as natural as possible. Yet these ingredients are essential to create the traditional flavors of New Orleans, where this dish is served in both homes and restaurants on Monday, laundry day, when there is less time for cooking; the beans were traditionally left on the stove to cook all day. You can make delicious beans and rice using all fresh ingredients, of course, but they won't have that essential Nola flavor.

There are nearly as many variations of red beans and rice as there cooks who make the dish and if you've grown up eating it, you don't need a recipe. You just cook it, getting to your destination by instinct gleaned from years of smelling and tasting. If you didn't grown up around red beans and rice, though, you do need to begin with a recipe, which you will likely alter as you learn your way around.

1 pound dried red beans, picked over to remove small rocks and other kinds of beans

4 or 5 large celery stalks, cut in fine dice

1 large or 2 medium green bell peppers, cut in fine dice

1 large or 2 medium white or yellow onions, peeled and diced

2 or 3 dried bay leaves

1 teaspoon garlic powder

1 teaspoon onion powder

1 teaspoon dried thyme (not ground)

1 teaspoon dried oregano (not ground)

1 teaspoon ground black pepper

1 teaspoon ground white pepper

¼ teaspoon ground cayenne pepper

2 teaspoons Tabasco sauce, plus more to serve

Kosher salt

2 cups uncooked long-grain white rice

Put the beans in a large container, add water to 2 inches above the beans and soak for 8 hours or overnight. Drain, rinse, and set aside.

Put the celery, bell peppers, onion, garlic and onion powder, thyme, oregano, peppers, and the 2 teaspoons of Tabasco sauce in a large saucepan or kettle, add 2 quarts (8 cups) of water, set over high heat, bring to a boil, reduce the heat and simmer for about half an hour. Skim off the foam and impurities that rise to the surface.

Add the beans and 2 teaspoons of kosher salt and continue cooking until the the beans begin to fall apart, about an hour and a half skimming off foam as needed. Stir frequently and add water as necessary to keep the mixture from becoming too thick; do not let the beans burn. When the beans are almost tender, cook the rice according to the package directions.

When the beans are done, remove from the heat. Taste and correct for salt.

To serve, place a generous scoop of rice in the center of individual serving plates and ladle the beans over the rice. Serve with Tabasco sauce on the side.

Variation:

- Add 1 or 2 smoked ham hocks along with the celery and other ingredients. When the beans are almost tender, remove the ham hocks, cool until easy to handle, pull the meat from the bones, and return to the pot.

- Add 1 pound andouille sausage, sliced into rounds and sautéed and either served on top of the beans after they are ladled onto the rice or stirred into the beans beforehand.

Black Pepper Polenta Made in a Slow Cooker

Makes 12 to 16 servings

When I wrote Polenta *(Broadway Books, 1997), I had never used a slow cooker or crock pot and it never occurred to me to try preparing polenta in one. But about a decade later, I invited twenty or so culinary students to my house for an early morning breakfast before we headed out on a field trip. I wanted to offer creamy polenta with an array of toppings but did not want to rise at 5 a.m. to prepare it. By then I had fallen in love with my slower cooker and decided to give it a try. I slept a tad nervously, wondering if it would turn out or if I'd have to face twenty hungry students with nothing to feed them.*

The polenta, perfectly creamy and full of true corn flavor, could not have been better. Soon thereafter, I offered the recipe in one of my weekly columns. Since it appeared, it has become my second most requested recipe, right after Preserved Lemons, which you'll find on page 281.

12 cups hot water
3 cups polenta, as fresh as possible
1 tablespoon kosher salt, plus more to taste
6 tablespoons butter, preferably organic
Black pepper in a mill
8 ounces grated cheese, Vella Dry Jack, Parmigiano-Reggian, Pepato, or
   Pecorino-Romano or a mixture of these or similar cheeses
Toppings of choice

Pour the water into a large slow cooker and set the control to high. Pour in the polenta, whisking to encourage the grains to separate. Add the salt, stir and cover.

Give the polenta a quick stir every 15 minutes or so until it begins to thicken, which will take about an hour and a half. Reduce the heat to low and cook for 4 to 5 hours, stirring now and then when you think of it. When the polenta is tender and creamy, season very generously with black pepper, reset the heat to warm and hold for up to 10 hours. If at any point the polenta seems too thick, thin it with a little water.

To serve, return the heat to low, stir in the butter and cheese, taste and correct for salt, and add several more turns of black pepper. Set condiments of choice alongside and let guests select what they like after ladling their polenta into a bowl.

Three-Peppercorn Bread with Pancetta & Garlic

Makes 1 loaf

You don't have to be an expert to make good bread at home; you don't even need to have a lot of time on your hands. This modest dough, which can be used for pizza and bread sticks as well as for a loaf, is simple and forgiving; I make it when I crave the sensual rhythm of kneading as a relief from a hectic schedule. Think of it as BLT bread with the bacon on the inside; the first time I made it, I sliced it while it was still warm, spread it with mayonnaise, and topped it with fresh summer tomatoes, coarse salt, and black pepper. Nothing could be better.

2 teaspoons yeast
⅓ cup warm water
3 ½ cups all-purpose flour
2 tablespoons crushed black peppercorns
1 tablespoon crushed white peppercorns
1 tablespoon crushed green peppercorns
2 teaspoons kosher salt
4 tablespoons extra virgin olive oil
4 ounces pancetta, minced
4 garlic cloves, minced
Cornmeal

In a large bowl, combine the yeast and warm water and set aside for 10 minutes. Using a whisk, stir in 1 cup water, 1 cup of the flour, all but 1 teaspoon each of the peppercorns, the salt, and 2 tablespoons of the olive oil. Switch to a heavy wooden spoon and add as much of the remaining flour, about half a cup at a time, as the dough will take. Turn the dough (it will be sticky) onto a floured surface and knead it gently until it is smooth and velvety, about 7 or 8 minutes. Wash and dry the mixing bowl, rub it with olive oil, and place the dough in the bowl. Cover with a tea towel and let the dough rise until it has doubled in size, about 2 ½ hours.

Turn the dough onto a lightly floured work surface and let it rest for a few minutes. Meanwhile, put 1 tablespoon of the remaining olive oil in a sauté pan, add the pancetta, and fry until it is almost but not quite crisp. Add the garlic, sauté for 1 minute more, and remove from the heat. Add the remaining tablespoon of olive oil and the remaining 3 teaspoons of crushed peppercorns.

Roll the dough into a rectangle about ¾ to 1 inch thick and spread the pancetta mixture over the dough's entire surface. Roll the dough into a log and tuck the ends under. Sprinkle a pizza paddle or the bottom of a baking sheet with cornmeal and let the loaf rise, lightly covered, until it has doubled in size (1 to 2 hours, depending on the temperature of the room).

About 30 or even 45 minutes before you plan to bake the loaf, place a baking tile or stone in the oven and preheat it at 400 degrees. Just before baking, scatter cornmeal over the surface of the stone and then carefully place the loaf on top of the stone. Reduce the heat to 350 degrees and bake for 35 and 40 minutes, until the crust is golden brown. Remove the loaf from the oven and cool on a rack for at least 15 minutes before slicing.

Variations:

To make bread sticks, let the dough rest after its first rise. Preheat the oven and two baking tiles to 375 degrees. Cut the dough in half, then cut each half in half again, and continue until you have 18 equal pieces of dough, Lightly flour your hands and roll each piece between your palms until it forms a rope about 8 to 10 inches long. Set each rope on a floured surface. Mix an egg white with a tablespoon of water and brush each bread stick with the mixture. Sprinkle each bread stick lightly with coarse salt (such as sel gris, Cyprus black salt, or Hawaiian red salt), set on baking tiles sprinkled with cornmeal, and bake until lightly browned, about 12 minutes. Cool on a rack and use within a day or two.

Pepper-Crusted Pizza with Porcini, Fontina & Sage

Makes 1 pizza

The recipe for dough (page 191) makes enough for two pizzas. I usually make one pizza, and use the other half of the dough to make spicy bread sticks (see the Variation at the end of that recipe). You'll have the best results if you bake pizza (or any bread) on a baking stone; you needn't buy an expensive one—unglazed Mexican paver tiles work well and cost about two dollars apiece or less. The highly prized porcini can be found in many U.S. farmers markets and specialty stores, but if you can't get them, use chanterelles or, in a pinch, criminis.

Three-Peppercorn Bread dough (page 191), taken through the first rise
¼ cup clarified butter
1 shallot, minced
3 or 4 fresh medium porcini, cleaned
2 teaspoons snipped chives
Kosher salt
Black pepper in a mill
4 ounces Italian Fontina, thinly sliced
8 to 10 fresh sage leaves
Cornmeal

Cut the dough in half and let it rest, covered, on a work surface.

In a medium sauté pan, heat 3 tablespoons of the butter over low heat until it turns golden brown and begins to give off a nutty smell. Cool slightly, add the shallot, and sauté for 7 to 8 minutes, until soft and fragrant.

Cut the porcini lengthwise into ¼-inch-thick slices; add them to the pan with the shallot, and sauté 3 to 4 minutes, turn, and sauté for 3 or 4 minutes more, until golden brown. Add the chives and season with salt and pepper.

Preheat the baking stone and oven at 475 degrees for at least 30 minutes before you bake the pizza.

Lightly flour a baker's paddle or other work surface and use your hands to shape one half of the dough into a 10-inch round. Melt the remaining tablespoon of butter, brush it over the surface of the pizza shell, and top with the cheese, arranging it in a single, overlapping layer. Arrange the porcini over the top of the cheese, spoon any pan drippings over the mushrooms, and add the sage leaves on top. Season with salt and several generous turns of black pepper.

Sprinkle the stone lightly with cornmeal and set the pizza on top. Bake for 10 to 12 minutes, until the cheese is bubbly and the crust is lightly browned.

Let the pizza cool for 3 or 4 minutes before cutting in wedges and serving.

Savory Zucchini Galette

Serve 4 to 6

Galette dough is the easiest and most forgiving dough, one I use to calm both students and friends who are nervous about the process. It can be used for both sweet and savory dishes. This dough can be used for any savory tart—tomato, eggplant, winter squash, onion, potato—and even for sweet galettes, especially strawberry, if you add a teaspoon of sugar to the dough and sprinkle the edges with sugar as well as salt and pepper.

Dough for 1 galette (recipe follows)
1 medium Romanesco zucchini, trimmed and cut into ⅛ inch rounds
Kosher salt
3 tablespoons butter, melted
Black pepper in a mill
3 ounces grated cheese, such as Laura Chenel Tome, Achadinha Capricious, or similar cheese
6 to 8 large basil leaves, stacked and shredded

First, make the dough and chill it.

Preheat the oven to 400 degrees.

Put the zucchini into a colander, sprinkle it generously with salt and let rest 30 to 45 minutes. Rinse the zucchini, turn it out onto a tea towel and pat it dry thoroughly.

Roll out the galette as described in the galette dough recipe and set it on top of a sheet pan lined with parchment. Brush the dough with butter, using about 2/3 of it. Arrange the zucchini on top, in concentric circles that overlap just slightly; leave about 2 inches of the outer edge of dough uncovered. Season the zucchini with salt and several very generous turns of pepper. Top with the cheese.

Gently fold the edges of the tart up and over the zucchini, pleating the edges as you fold them. Use a pastry brush to coat the edge of the tart with the remaining butter; season the edge with salt and pepper.

Bake until the pastry is golden brown and the zucchini tender and fragrant, about 35 to 40 minutes.

Transfer to a rack to cool, cut into wedges, top with the shredded basil and serve warm.

Savory Galette Dough

Makes 2 large or 8 small Galettes

2 cups (8 ounces) all-purpose flour
¾ teaspoon kosher salt
1 teaspoon freshly ground black pepper
6 ounces unsalted butter, cold
½ cup ice-cold water

In a medium bowl, combine the flour, salt, and the pepper, if using. Cut in the butter, using your fingers or a pastry cutter, until the mixture resembles coarse cornmeal; work very quickly so that the butter does not become too warm. Add the ice water and press the dough gently until it just comes together; do not overmix—it's okay if there appears to be unmoistened flour. Spread a sheet of plastic wrap over a flat surface and turn the dough out onto it.

Grip the ends of the plastic wrap and pull them together, so that the wrap presses the dough together. Wrap the dough into a ball and refrigerate it for at least 30 minutes. (At this point, the dough can be wrapped a second time and stored in the freezer for up to 3 months.)

To make the galette, cut the dough into 2 equal pieces; wrap one and return it to the refrigerator.

Set the other piece of dough on a floured work surface and use the palm of your hand to pat it flat. Using a rolling pin, roll the dough into a circle about ⅛-inch thick and about 14 inches in diameter.

Set the dough on a baking sheet covered with parchment paper and keep chilled until ready to fill. (The dough can also be frozen after it has been rolled; be sure to wrap it tightly.)

Flesh

Salt and pepper bring the meats we eat to life, lifting a shrimp, a chicken thigh, a juicy steak to its full potential. Some foods—rack of Sonoma lamb, for example—need nothing more than a bit of both. Other foods—corned beef, example—rely on the effect of salt to transform them, and some, such as bratwurst, would be pale shadows of themselves without pepper.

In this chapter, I offer a selection of dishes from different parts of the world, some, such as Nana's Pot Roast, from my earliest and most formative years, and others, such as Black Pepper Crab, that I savored during my explorations of salt and pepper.

Nola's Famous Barbecue Shrimp

Salt & Pepper Shrimp

Black Pepper Crab

Dry-Roasted Mussels with Black Pepper, Roasted Sweet Peppers & Hot Vinegar

Wild Pacific King Salmon with Sweet & Sour Peppercorn Sauce

Chicken Alla Diavola

Ginger Pepper Duck Breast with Cherry-Basil Sauce

Rack of Lamb with Spring Onions & Potato Purée

Spicy Pork Chops with Sarawak Sambal

Kalua Pig

Kalua-Style Turkey Thighs with Sweet Potatoes

Steak au Poivre Blanc

Steak au Poivre Rouge

Steak au Poivre Vert

Nana's Pot Roast

Three-Peppercorn Meatloaf

Classic Corned Beef and Cabbage

Standing Rib Roast with Carrots in Peppercorn Cream

Bratwurst with Sauerkraut & Apples

A Contemporary Haggis with Neeps & Tatties

A Salt & Pepper Cookbook

Nola's Famous Barbecue Shrimp

Serves 2 as an entree, 4 to 6 as an appetizer

The barbecue shrimp of New Orleans are like no other barbecue anything I've ever encountered. There's no barbecue sauce as we think of it, there's no grill, no wood or charcoal heat. It is typically made inside, not outside. It is a thing unto itself and surprisingly easy to duplicate at home. Serve it directly in the pan in the center of the table as an appetizer, spoon it over creamy grits (or polenta) or simply as it is, with plenty good bread for sopping up the irresistible sauce.

If you love New Orleans–style foods, it is worth it to make your own Creole seasoning, though there are pretty decent commercial ones, too.

2 pounds large wild Gulf shrimp, heads and shells intact
6 ounces (1 ½ sticks) organic butter, cut into small cubes, chilled
1 lemon, cut into 6 lengthwise wedges
3 garlic cloves, crushed and minced
½ cup Crystal hot sauce
⅓ cup Worcestershire sauce
Juice of 1 lemon
1 teaspoon ground black pepper
2 teaspoons cracked black pepper
1 tablespoon Creole seasoning (page 378)
Kosher salt
2 tablespoons heavy cream
Sourdough hearth bread

Rinse the shrimp in cool water, set on a clean tea towel, pat dry and set aside.

But a generous tablespoon or so of butter in a heavy skillet, preferably cast iron, set over medium heat and when the butter is foamy, add the lemon wedges. Sauté on one side for about 2 minutes, until the wedges have picked up a bit of color. Turn, cook 2 minutes more and transfer to a bowl or plate.

Add the garlic, sauté for 1 minute and add the hot sauce, Worcestershire sauce, lemon juice, peppers, and Creole seasoning. Stir and simmer on fairly high heat until the sauce begins to thicken a bit, about 3 to 4 minutes.

Add the shrimp, season with salt, cook for 1 minute, turn and cook for 2 minutes more. Reduce the heat to low and begin to add the butter, about a tablespoon at a time, swirling the pan until it melts before adding more. Add the cream, return the lemon wedges to the pan, toss, remove from the heat and serve right away, with hot bread alongside.

Salt & Pepper Shrimp

Serves 4 to 6

Shrimp and other small crustaceans cooked with spices are common throughout Asia as well as in the American South. There seem to be as many versions of this dish as there are cooks. Some coat the shrimp in a batter before cooking them, others don't, and many use either chiles or Sichuan peppercorns instead of black pepper. This version, which I can't blame on anyone but myself, features the full range of heat and flavor from several of the most common peppers and chiles in the world: back and white peppercorns, Sichuan peppercorns, serrano chiles, and chipotles. It is more Asian than American, its influences are Thai as well as Chinese, and I encourage you to fiddle with my version as much as I have with the originals. The main things to keep in mind are that the shrimp must be cooked in their shells (split the back of the shells to devein them) and that the heat must be high; otherwise, the shrimp will become dry and hard.

Dipping Sauce (recipe follows)
2 pounds medium shrimp in their shells, washed, deveined, and dried on a
    tea towel
¼ cup dry Marsala, sherry, or mirin
2 teaspoons kosher salt, plus more as needed
2 teaspoons Sichuan peppercorns, toasted, crushed, and sifted (see Note,
    page 271)
1 teaspoon ground black pepper
1 teaspoon ground white pepper
1 teaspoon chipotle powder
6 garlic cloves, minced
2 serrano chiles, minced
One 1-inch piece ginger, peeled and diced
2 cups peanut oil
8 to 10 scallions, white and green parts, cut into thin rounds
6 cups shredded iceberg lettuce

First, make the dipping sauce and set it aside.

Place the shrimp in a bowl, drizzle the Marsala over them, and toss them lightly. Sprinkle with the 2 teaspoons of salt, the Sichuan peppercorns, black pepper, white pepper, and chipotle powder, toss again, and set aside for 10 minutes.

Use a suribachi to grind the garlic, serranos, ginger, and a generous pinch of salt to a paste. Drain the shrimp and discard any liquid that has collected in the bowl. Pour the oil into a wok, set over high heat, and when hot, add the shrimp. Cook for 1 minute, tossing and stirring constantly, until the shrimp begin to turn opaque pink. Set a strainer over a bowl, carefully pour in the shrimp and the oil, and lift the strainer to drain the shrimp.

Return 2 tablespoons of the cooking oil to the wok, return to high heat, add the garlic mixture, and cook, stirring constantly, for 30 seconds. Add the shrimp and half of the scallions, toss, and cook for 1 minute. Remove from the heat.

Spread the lettuce on a large serving platter, top with the shrimp and cooked scallions, and sprinkle with the remaining uncooked scallions. Season with a little salt and several turns of black pepper, and serve with the dipping sauce.

for the dipping sauce
¼ cup fresh lime juice (from 1 to 2 limes)
¼ cup rice vinegar
2 tablespoons sugar
1 teaspoon kosher salt
3 garlic cloves, minced
2 serrano chiles, minced, or 1 teaspoon chipotle flakes
2 teaspoons fresh grated ginger
2 tablespoons minced fresh cilantro leaves

Combine the lime juice, vinegar, sugar, and salt in a small bowl and stir until the sugar and salt are dissolved. Add the garlic, serranos, ginger, and cilantro leaves, stir, and set aside until ready to serve.

Makes about ½ cup.

Black Pepper Crab

Serves 3 to 4

San Pedro's Cafe is a small restaurant near the Portuguese Square, in the Portuguese settlement, a historic community in Malacca in West Malaysia, about three hours south of Kuala Lumpur. San Pedro's specializes in the Portuguese-influenced foods of the region, and serves crab half a dozen ways, including this way, one of the rare Malaysian dishes that includes a substantial amount of black pepper.

¼ cup clarified butter
3 shallots, minced
8 garlic cloves minced
1 serrano chile, minced
2 tablespoons dried shrimp, ground
2 tablespoons freshly cracked black peppercorns
2 tablespoons dark soy sauce
2 tablespoons oyster sauce
2 tablespoons brown sugar
2 large Dungeness crabs, cooked, cleaned, and broken apart

In a wok or sauté pan, heat the clarified butter, add the shallots and sauté for 3 or 4 minutes, add the garlic and serrano, and sauté for 2 minutes more. Do not let the ingredients brown.

Stir in the shrimp, peppercorns, soy sauce, oyster sauce, and brown sugar, and toss. Add the crab, toss to coat thoroughly with the butter mixture, add ½ cup water, cover, and simmer for 7 to 8 minutes, until the crab is heated through. Transfer to a platter and serve immediately, with plenty of napkins for wiping the rich juices off your fingers.

Dry-Roasted Mussels with Black Pepper, Roasted Sweet Peppers & Hot Vinegar

Serves 4

Bistro Cooking by Patricia Wells (Workman Publishing, 1989) is one of my favorite cookbooks and it is that treasured tome that I discovered her simple recipe for "Burn Your Fingers Mussels," mussels that are simply grilled dry until they pop open and then seasoned with a lot of freshly ground black pepper. They are so very good. This recipe, slightly more elaborate but still quite simple, is inspired by Patricia's.

Habanero Vinegar (see Note on next page) or other spicy vinegar
6 or 7 red sweet peppers
2 tablespoons red wine vinegar
5 to 6 garlic cloves, slivered
3 pounds fresh PEI mussels, rinsed in cool water
Black pepper in a mill

If you do not have spicy vinegar, make the Habanero Vinegar the day before preparing the mussels.

Roast the bell peppers over a high flame until the skins are blistered. Put the seared peppers in a bowl, cover with a tea towel and let cool.

Use you fingers to remove the skins; use a pairing knife to cut out the stem and seed core.

Set a pepper on a work surface and cut across the bottom (blossom end) and up one side, so that the pepper can be opened into a rectangle; continue until all the peppers are similarly cut. Cut the peppers into medium julienne and spread them over a large platter. Scatter the garlic on top, drizzle with olive oil and use your fingers or two forks to gently mix together. Set aside.

Set a large heavy skillet over high heat for 2 minutes. Sprinkle a little water in the pan and if it sizzles immediately, it is ready; if it does not, heat another minute or so and test again.

When the pan is sufficiently hot, add as many mussels as the pan can hold in a single layer. The mussels will begin to pop open almost immediately. Shake the pan gently until all the mussels pop open. Working quickly, use a slotted spoon to transfer the cooked mussels to the platter with the roasted peppers. Continue until all the mussels have been cooked.

Grind a generous amount of black pepper over the mussels and peppers. Serve immediately, with the vinegar alongside to use as a condiment.

NOTE

Make several lengthwise cuts in a habanero or other hot pepper, put it in a bowl and pour a cup of champagne or white wine vinegar over it. Cover lightly with a tea towel or cheese cloth and let sit for at least 12 hours and as long as 36 hours. Remove and discard the habanero, put the vinegar into a narrow bottle and cover. It will keep indefinitely.

A Salt & Pepper Cookbook

Wild Pacific King Salmon with Sweet & Sour Peppercorn Sauce

Serves 4

Slightly sweet, a little hot, and pleasantly tangy, this sauce provides a creamy and inviting coating for the sweet salmon. If you have it, use the red Hawaiian alae salt to finish this dish; it's the same color as the salmon and will add a beautiful flourish.

1 teaspoon white peppercorns
1 teaspoon black peppercorns
1 tablespoon dried green peppercorns
1 teaspoon kosher salt, plus more to taste
4 wild Pacific King salmon fillets, about 6 ounces each, pin bones removed

3 to 4 tablespoons unsalted butter, chilled
1 teaspoon sugar
1 tablespoon green peppercorns in brine, drained
½ cup fruity white wine (such as Viognier)
Juice of ½ lemon
Hawaiian alae salt

Using a suribachi or heavy mortar and pestle, grind the white and black peppercorns to a medium coarseness. Add the green peppercorns and grind the mixture together until all the peppercorns are uniformly ground to a medium-fine coarseness; add 1 teaspoon kosher salt and mix together. Set the salmon fillets on a work surface and use your fingers to sprinkle the pepper mixture over the entire surface of the salmon, pressing lightly to make it stick.

Melt 1 to 2 tablespoons of the butter in a heavy sauté pan over medium heat, add the fillets skin-side up, and cook until golden brown, about 2 to 3 minutes. Turn the fillets skin-side down, reduce the heat to low, cover, and cook up to 5 minutes. Remove the lid and cook until done, about 3 to 5 minutes more, depending on the thickness of the fillets. Transfer the fillets to a serving platter and keep warm.

Increase the heat to medium, add the sugar, brined peppercorns, and wine, and simmer until the wine is reduced by about ⅔ Add the lemon juice and 1 tablespoon of the remaining butter. Swirl the pan as the butter melts, but do not let it boil. Add the remaining butter, swirl until it melts, taste, and add a generous pinch of salt to balance the flavors. Pour the sauce over the salmon, sprinkle each serving with a little Hawaiian salt, and serve immediately.

Chicken Alla Diavola

Serves 4

You find pollo alla diavola *in traditional Italian restaurants, those homey, turn-of-the-century eateries often in a basement and frequently decorated with photographs of famous Italian actors and politicians on the walls. Sometimes, the dish includes mustard, a French rather than Italian addition. (For a delicious Asian twist, see the Variation at the end of the recipe.)*

Please note that the chicken is rinsed to remove any lingering liquids that might influence the taste, not to remove bacteria from the chicken. Obviously, when a chicken is washed under running water, there should be nothing—no salad greens, no other ingredients whatsoever—underneath the flow of the water to become contaminated by the run-off.

1 large pastured chicken, cut into pieces
3 tablespoons black peppercorns, coarsely crushed
2 to 3 teaspoons kosher salt
⅓ cup fresh lemon juice
½ cup extra virgin olive oil
2 lemons, cut into wedges

Rinse the chicken pieces under cool water and dry them with a tea towel. Place the chicken in a single layer in a glass baking dish.

Sprinkle the black pepper over the chicken, turning so that both sides are coated. Sprinkle 2 teaspoons of the salt over the chicken, then drizzle with the lemon juice and the olive oil. Cover and let the chicken marinate in the refrigerator for 4 hours or overnight, basting occasionally.

To cook the chicken, prepare a charcoal fire. When the fire is ready, set the rack about 5 inches from the fire, set the chicken skin-side up on the rack, and broil for 15 minutes, or until the chicken is lightly brown. Baste the skin side with the marinade (for an added element of flavor, use large stalks of rosemary or sage for the basting), turn, and baste the cooked side.

Continue to grill, rotating the chicken to mark it evenly, and basting occasionally. Turn a final time and cook until the juices run clear, for a total of 25 to 30 minutes. Transfer the bird to a serving platter, let it rest for 5 to 10 minutes, season with the remaining teaspoon of salt, garnish with lemon wedges, and serve immediately.

Variation:

Add a tablespoon of minced fresh ginger and a teaspoon of crushed red pepper to the olive oil before drizzling it over the chicken.

A Salt & Pepper Cookbook

Ginger Pepper Duck Breast with Cherry-Basil Sauce

Serves 4 to 6

The breast from the plump Muscovy duck, used in the California to make foie gras until the state banned the practice in 2012, is as thick and tasty as a good steak and lends itself to similar preparation. In this version, ginger and garlic contribute spicy, aromatic elements and cherries contribute a sweet tang.

1 whole Muscovy duck breast
One 1-inch piece fresh ginger, peeled and chopped
2 garlic cloves
1 teaspoon kosher salt
2 teaspoons freshly crushed black peppercorns
1 teaspoon freshly crushed white peppercorns
1 teaspoon freshly crushed dried green peppercorns
1 teaspoon Sichuan peppercorns, toasted, crushed, and sifted (see Note, page 271)
½ teaspoon allspice, crushed
2 tablespoons butter, preferably organic
1 shallot, minced
1 cup firm-ripe cherries (Bing or a mix of Bing and Queen Anne), pitted and cut in half
Olive oil
½ cup dry Marsala
7 to 8 fresh basil leaves, shredded

Cut the duck breast in half down the center, separating the two halves. Using a small sharp knife, remove the fat, which will come off easily in most places. Leave the skin in place but trim it so that it does not hang over the meat; score the skin diagonally, with cuts about 1 inch apart and then score in the other direction, creating a patchwork of squares. Reserve the fat for another purpose and set the breasts on your work surface.

Using a molcajete or suribachi, grind the ginger, garlic, and salt into a fine paste. Stir in the peppercorns and allspice, and use your fingers to rub the paste into the entire surface of each piece of meat. Cover and let sit for at least 1 hour or overnight in the refrigerator.

Just before cooking the duck, put the butter into a small sauté pan and when it is melted, add the shallot and sauté until soft and fragrant, about 7 minutes. Season with salt and pepper. Add the cherries, toss to coat with butter, heat through, remove from the heat and set aside, covered.

Brush a ridged cast-iron skillet very lightly with olive oil and set it over medium-high heat. When the pan is very hot, add the duck breast and cook, skin-side down, rotating once to mark the breasts, for about 6 minutes for 1-inch-thick breasts (less for thinner meat). Turn the duck over, mark on the ridges, rotate, and add the Marsala, pouring it over the duck. Cook for 3 to 4 minutes more for rare meat, 7 to 8 minutes for medium rare.

Remove from the heat, cover loosely and let the duck rest for 5 minutes before cutting into ¼-inch diagonal slices.

While the ducks rests, set the pan with the cherries over medium heat, add the remaining butter and swirl as it melts. Remove from the heat, correct the seasoning and add the basil.

Arrange on individual plates, spoon the cherries and the pan juices over the duck and serve immediately, with accompaniments of choice.

Rack of Lamb with Spring Onions & Potato Purée

Serves 2

Few meals demonstrate, at least to me, the transformative powers of salt and pepper. When you have all the ingredients at hand, you can have this feast on the table in under 45 minutes. You need no elaborate sauces or complicated technique, just good ingredients seasoned properly. Sometimes I add Roasted Asparagus (page 288) to the plate and, if I want to draw out the meal, start with Radishes with Butter & Salt (page 91) and end with a piece of season fruit and, maybe, a peppery cheese. If you love wine, this is a meal that calls for the very best, most delicate Pinot Noir you can find.

1 rack of young lamb, 5 or 6 ribs, Frenched
Flake salt
Black pepper in a mill
¾ pound potatoes, such as German Butterball, sliced
6 small spring onions, trimmed and sliced in half lengthwise
4 ounces best-quality organic butter, cut in pieces, chilled
¼ cup heavy cream, hot
1 tablespoon freshly snipped chives
1 chive flower, little blossoms separated
Fleur de sel

Season the meat, including between the ribs, all over with salt and pepper. Set aside and cover with a sheet of wax paper.

Preheat the oven to 300 degrees.

Put the potatoes into a medium sauce, cover with at least an inch of water, season generously with salt and bring to a boil over high heat. Lower the heat and simmer until just tender, about 10 to 12 minutes. Drain thoroughly, return to the heat and cook off any remaining water. Remove from the heat, cover and keep hot.

Set a ridged cast-iron pan over high heat and when it is hot, sear the fat side of the rack of lamb until it begins to sizzle and brown a bit. Turn the meat fat side up, add the spring onions alongside and set on the middle rack of the oven for 20 minutes. Test the temperature and if it has not reached 120 degrees, return it to the oven for 10 to 15 minutes more.

Remove from the oven and cover the lamb and onions with aluminum foil or a domed lid.

Set the potatoes over medium-low heat and whip in the butter, a piece at a time. Add the hot cream, taste, season with salt and several very generous turns of black pepper.

Working quickly, divide the potatoes between two pretty dinner plates. Cut the rack into individual chops and set them, their bones resting on the potatoes, alongside. Add the spring onions, season everything with black pepper and sprinkle the lamb with fleur de sel. Garnish the potatoes with chives and chive flowers and serve immediately.

Spicy Pork Chops with Sarawak Sambal

Serves 4

As I gathered, developed, and tested recipes for this book, I invited friends over to taste the results at the end of the day. I served this simple dish, along with about a dozen other, more complex dishes, so that it would not go to waste, as I had decided not to include it. By the end of the evening, I was overruled; it was the favorite of the night.

Sambal is a savory condiment, typically with a fair amount of heat, used in Southeast Asia and India.

4 to 6 pork loin chops, about 6 ounces each

1 tablespoon freshly cracked black pepper

1 teaspoon dried green peppercorns, cracked

¼ cup white vinegar (see Note)

Kosher salt

4 tablespoons Sarawak Sambal (page 361)

Set the loin chops on a plate or in a shallow dish that holds them in a single layer. Mix the peppers together and sprinkle over the chops, covering them lightly but entirely. Drizzle the vinegar over the chops, cover the dish with plastic wrap, and let it sit for at least 1 hour or up to 6 hours in the refrigerator.

Warm the chops to room temperature 30 minutes before cooking them. Heat a stove-top grill or ridged cast-iron pan and when it is hot, cook the pork chops, rotating once to mark them, for 7 to 8 minutes, until they are lightly browned. Turn them and cook, again rotating once, until just done, an additional 5 to 7 minutes depending on the thickness of the meat.

Set the chops on a work surface, cut the meat on the diagonal into thin slices, arrange on individual plates, and add a spoonful of sambal to each serving.

NOTE

Use Vinaigre de Banyuls, sherry vinegar, or a champagne vinegar. In a pinch, any white vinegar will do, though keep in mind that distilled vinegar has a sharp taste that can dominate a dish.

Kalua Pig

Serves 8 to 10

Few aromas are more deliciously seductive than that of smoky pork as it is uncovered and lifted from the underground oven known as an imu. It is the heart of Hawaiian entertaining, of the luau, familiar to most people as a somewhat kitschy extravaganza staged in resort hotels but, in homes throughout the Hawaiian Islands and beyond, wherever Hawaiians live, it is a traditional and beloved celebration. It is often a potluck, with each family member bringing their specialty. When no one cooked the entire feast, no one had to suffer hours of heat. It makes sense in a warm climate.

It is easy to make great Kalua Pig in an oven and over the years, I've perfected this easy and pretty much foolproof technique. It should always be accompanied by steamed white rice, mac salad or potato-mac salad, and several other Hawaiian dishes, many hard to duplicate without access to traditional ingredients such as taro leaves for squid luau and laulau, taro root for poi, and coconut milk for haupia. I add Pineapple with Black Pepper (page 101) alongside and love to sprinkle Hawaiian Chile Water on both the pork and the rice.

Hawaiian alaea salt is traditional, too, and should always be offered at a Hawaiian-style feast.

Sometimes I depart from Hawaiian tradition and serve both lime wedges and hot corn tortillas with this pork, especially the leftovers, which make great tacos.

Large (about 6 pounds) pork butt (Boston butt) or pork shoulder
Liquid smoke
4 tablespoons kosher salt
Hawaiian alaea salt
2 limes, cut in wedges, optional
Black pepper in a mill

If you have a clay roaster, prepare it according to the manufacturer's instructions.

Preheat the oven to 275 degrees.

Set the pork roast on a clean work surface and score the fat layer with deep diagonal cuts about 1 inch apart; make another row of cuts at right angles to the first cuts. Use a pastry brush to paint the pork with a layer of liquid smoke, making sure to brush into the cuts. Rub kosher salt into the pork, including into the cuts, and be sure to use all of it.

Set the pork in a clay roaster or deep roasting pan. Pour about ¼ inch of water into the pan. Cover with the lid or seal tightly with aluminum foil.

Set in the oven and cook for 5 hours, or until the pork falls apart when pressed.

Transfer the pork in its pot to a work surface, leave covered and let rest 15 to 30 minutes.

Carefully transfer the pork from the pot to a large serving platter. Use two forks to shred the pork. Sprinkle lightly with Hawaiian salt, garnish with lime wedges, if using, and serve, with black pepper alongside.

Kalua-Style Turkey Thighs with Sweet Potatoes

Serves 6 to 8

I developed this recipe after listening to friends discuss making their Thanksgiving turkey kalua-style. Thanksgiving had passed and I didn't want to wait a year to give it a try and so I used turkey thighs, which hold up so much better to lengthy cooking that turkey breast. It's easy, nearly foolproof and makes a wonderful alternative to Kalua Pig when you're eating with friends who don't eat pork.

2 to 3 bone-in skin-on turkey thighs
Liquid smoke
Hawaiian Alaea salt
6 ti leaves, central stalk removed, optional
3 large sweet potatoes
Black pepper in a mill
Steamed rice
Hawaiian Chile Water (page 363)

Preheat the oven to 400 degrees.

Set the turkey thighs on a clean work surface.

Pour about half a bottle of liquid smoke into a wide bowl, add the turkey, turn to fully coat it and cover lightly with a tea towel. Set aside for 15 to 30 minutes.

Transfer the turkey thighs to a work surface and rub them all over with salt.

If using ti leaves, wrap each thigh in two leaves. If not using ti leaves, wrap each thigh in a large sheet of parchment and tie it closed, somewhat loosely, with kitchen twine.

Set a large sheet of aluminum foil on a clean flat surface and top it with a second sheet placed perpendicular to the first. Set the turkey thighs on top and wrap them loosely, first with the top sheet of foil and then with the bottom sheet. You want a closed package but you do not want to wrap the turkey too tightly.

Set a rack on a baking sheet, set the turkey on the rack and then set on the middle rack of the oven. Cook for 30 minutes, reduce the heat to 225 degrees and cook for 3 to 4 hours more, until the turkey is very tender.

Meanwhile, peel the sweet potatoes and cut them into thick (about ⅜-inch) rounds. Set them in a steamer basket over boiling water and steam until nearly but not quite done. Remove from the steamer.

Spread the sweet potatoes over a nonstick baking sheet or pan, season lightly with salt and generously with black pepper, and set aside.

Remove the turkey from the oven, increase the heat to 350 degrees and set the sweet potatoes on the middle rack. Cook, turning once, until fully tender, about 15 minutes.

Unwrap the turkey carefully and remove the ti leaves, if using. Carefully tip the juices that have collected into a small bowl. Use two forks to shred the turkey.

Pile the roasted sweet potatoes on one side of a large platter. Set the shredded turkey alongside and drizzle juices over both. Season all over with black pepper and serve, with rice, cooking juices, steamed rice, and chile water alongside.

Variation:

Trim and core a small green cabbage and cut it into very thin slices. When the turkey is done, tip the cooking juices into a sauté pan, add the cabbage, cover the pan and sauté, stirring once or twice, until the cabbage is tender. Serve on the same platter as the turkey and sweet potatoes.

Steak au Poivre Blanc

Serves 2, easily doubled

"It was my favorite seduction dinner when I was single," a friend told me when I mentioned steak au poivre, "and it always worked." Whether you are courting Aphrodite or merely fixing dinner, the many variations of steak au poivre will rarely if ever let you down. In this version, white peppercorns, whose flavors blossom with high heat, are the key ingredient. The result is a sultry dish with a great depth of flavor. If seduction is your goal, don't skimp on the red wine and don't overeat (or overfeed your guest); you want a little hunger to linger after the meal is over.

2 tablespoons white peppercorns
1 tablespoon black peppercorns
2 teaspoons kosher salt
2 thick steaks (New York, rib-eye, or market steaks), preferably grass-fed,
    about 8 to 10 ounces each
Olive oil
½ cup dry Marsala
½ cup heavy cream
Half-and-half, as needed
Fleur de sel

Using a mortar and pestle (or a tea towel and rolling pin) crush the peppercorns to medium coarseness. Combine the crushed peppercorns and the salt in a small bowl. Using a pastry brush, brush each steak on both sides with a little olive oil. Using your fingers, cover both sides of each steak with the peppercorn mixture, pressing the pepper into the meat and coating it entirely. Set the steaks in a single layer on a plate or tray, cover with wax paper or plastic wrap, and refrigerate for at least 1 hour or up to 4 hours.

Preheat the oven to 200 degrees. Remove the steaks from the refrigerator and heat a large heavy skillet over high heat. When the skillet just begins to smoke, add the steaks. Cook on one side for 4 to 5 turn, and cook for 3 to 4 minutes more for rare steaks (6 to 7 minutes for medium rare).

Turn off the oven.

Set the steaks on a warm platter and set them in the warm oven. Reduce the heat under the pan to medium, add the Marsala, and deglaze the pan, using a whisk or wooden spoon to loosen pan drippings. When the Marsala has reduced to about 2 tablespoons, add the heavy cream, bring to a boil, and reduce by ⅓. If the sauce seems too thick, thin with half-and-half until you reach the desired consistency.

Set the steaks on individual plates, top with some of the sauce, sprinkle with fleur de sel and and serve immediately.

> ### *Variation:*
>
> Make a similar dish using Muscovy duck breast instead of steak. Let rest for 5 minutes after cooking and cut into ¼-inch thick diagonal slices. Spoon sauce onto individual serving plates and top with several slices of duck. Season with fleur de sel or sel gris and a turn or two of black pepper before serving. A single duck breast will serve 2 to 3.

Steak au Poivre Rouge

Serves 4

Black peppercorns are more aromatic than white peppercorns, and are perfect in concert with the red wine of the sauce. Serve with a simple green salad and roasted new potatoes.

4 tablespoons whole black
    peppercorns
1 tablespoon whole white
    peppercorns
1 tablespoon kosher salt
4 thick steaks (New York, rib-eye,
    or market steaks), about 8 to 10

    ounces each
Olive oil
1 ½ cups hearty red wine
2 tablespoons butter, cut into 4 pieces
    and chilled
1 tablespoon snipped chives

Using a mortar and pestle (or a tea towel and rolling pin) crush the peppercorns to medium coarseness. Combine the crushed peppercorns and the salt in a small bowl. Using a pastry brush, brush each steak on both sides with a little olive oil. Using your fingers, cover both sides of each steak with the peppercorn mixture, pressing the pepper into the steak. Set the steaks in a single layer on a plate or tray, cover with plastic wrap, and refrigerate for at least 1 hour or up to 4 hours.

Preheat the oven to 200 degrees. Remove the steaks from the refrigerator and heat a large heavy skillet over high heat. When the skillet just begins to smoke, add the steaks. Cook on one side for 4 to 5 minutes, turn, and cook for 3 to 4 minutes more for rare steaks (6 to 7 minutes for medium rare).

Turn off the oven. Set the steaks on a warm platter and place in the warm oven. Reduce the heat under the pan to medium, add the red wine to the hot skillet, and deglaze the pan, using a whisk or wooden spoon to loosen the pan drippings. Simmer until the wine is reduced by half. Quickly whisk in the butter, one piece at a time. Remove from the heat immediately after whisking in the last piece of butter. Taste and correct for salt.

Set steaks on individual plates, top with some of the sauce, scatter chives on top, add accompaniments, and serve immediately.

Steak au Poivre Vert

Serves 4

Let's change things up a bit, shall we? Steak au poivre is almost always made with a premium cut of beef. I prefer ribeye but New York steak works beautifully; some chefs favor the tenderness of beef tenderloin. You can easily make this dish with any of those cuts. But instead of always following tradition, sometimes I use skirt steak, which has the deliciously concentrated flavor. The secret to success is to not cook it beyond rare; if you do, it becomes tough.

Serve alongside Risotto Pepato (page 176) and Black Pepper Zucchini (page 293), if zucchini is in season. If not, simple wilted spinach is perfect alongside.

1 ¼ pounds skirt steak, cut into four equal pieces
Kosher salt
Mixed peppercorns in a mill
¼ cup brandy or Cognac
2 tablespoons butter
1 shallot, minced
2 tablespoons brined green peppercorns, drained
½ cup beef stock
¼ cup heavy cream
1 tablespoon snipped fresh chives
Sel gris or Maldon salt flakes

Season the steak all over with salt and pepper and press the pepper into the meat. Set aside for 15 to 45 minutes, while you assemble all the other ingredients, including side dishes. Have everything at the ready before you start cooking the steak.

Preheat the oven to 200 degrees.

Set a large heavy skillet over high heat and when it is very hot, add the steak. Cook for 90 seconds, turn and cook 90 second more. If the steaks are particularly thick, cook 2 minutes per side.

Quickly transfer to a plate and set in the warm oven.

Return the pan to the heat, add the brandy or Cognac and swirl until is nearly reduced. Add the butter and swirl until melted. Reduce the heat to medium, add the shallot and cook until soft and fragrant, about 7 minutes. Add the green peppercorns and the stock, increase the heat and simmer for 2 or 3 minutes so the sauce thickens. Add the cream, taste, correct for salt and remove from the heat.

Working quickly, cut the steaks in diagonal slices, set on warm plates where the side dishes are already in place, spoon sauce on top, sprinkle with chives and a few salt flakes and serve immediately.

Nana's Pot Roast

Serve 4 to 6

    I didn't know my grandmother very well and she died while I was a young girl. I remember her cooking as well as I remember her, and I regret that I didn't know her well enough to ask her questions in the kitchen. The dish I remember most clearly was her pot roast; the gravy had a wonderfully tangy flavor, the meat melted in my mouth, and the chunks of potatoes with their perfectly browned sides were the best I'd had back then. While I was experimenting with slow-cooked beef, inspired by a recipe in Laurie Colwin's More Home Cooking, *I came up with something that tasted surprisingly like my grandma's pot roast. The meat cries out for potatoes, so be sure to serve some alongside. Do not cook them in the pot with the meat, though; they'll soak up all the juices and both meat and potatoes will be dry.*

One 3 ½- to 4-pound chuck roast
2 teaspoons kosher salt
Black pepper in a mill
1 ½ cups beef stock or red wine
1 tablespoon kneaded butter (see Note below)

Preheat the oven to 200 degrees. Season the meat all over with salt and a generous amount of black pepper, place it in an ovenproof pot with a lid, add about ¼ inch of water, cover, and bake for 4 ½ to 5 hours, until the meat is fork tender. Remove the pot from the oven, transfer the meat to a serving dish, cover, and keep hot. Set the pot over a medium-high burner, add the wine or stock, stir with a whisk to loosen any bits of meat stuck to the pan, and simmer until the liquid is reduced by half.

Lower the heat, whisk in the kneaded butter a teaspoon at a time, and remove from the heat as soon as the last addition is incorporated into the sauce. Pour the gravy over the meat and serve immediately.

NOTE

To make kneaded butter, combine 1 tablespoon cool (not cold, but not too warm) butter with 1 tablespoon all-purpose flour. Use a fork to mix the ingredients together until smooth. Cover and refrigerate until ready to use.

Three-Peppercorn Meatloaf

Serve 6 to 8

This meatloaf may be a love-it-or-hate-it sort of thing. My feisty mother-in-law, a Dallas gal named Bess who remained a dear friend long after my relationship with her son concluded, typically loved my cooking. But when I visited her one day and said I was bringing over meatloaf and some other foods the next weekend, she said, "I love meatloaf but none of that stuff you make with all those damned peppercorns!"

The peppercorns in this meatloaf contribute both flavor and texture (the part Bess didn't like). Poaching keeps everything moist, and the mustard—both in the loaf itself and alongside—adds a tangy element. If you prefer to bake meatloaf, consult the Variation at the end of this recipe.

4 ounces pancetta, cut into small dice
2 tablespoons olive oil
1 yellow onion, diced
6 garlic cloves, minced
1 ½ pounds ground beef
¾ pound ground pork
3 tablespoons Dijon mustard
1 egg, beaten
¾ cup dried (not toasted) bread crumbs
¼ cup minced fresh flat-leaf parsley
1 tablespoon crushed dried green peppercorns
1 tablespoon crushed black peppercorns
2 teaspoons crushed white peppercorns
6 to 8 cups beef or veal stock
Green Peppercorn Mustard (page 358)

Fry the pancetta over medium heat until it is almost but not quite crisp. Add the olive oil and, when it is hot, add the onions, and sauté until they are limp and fragrant, about 8 minutes. Add the garlic, sauté for 2 minutes more, and remove from the heat. Let cool slightly.

In a large bowl, combine the onion mixture, beef, pork, mustard, egg, bread crumbs, parsley, peppercorns, and salt. (To check the seasoning, shape a small quantity of the mixture into a patty and sauté it on both sides, taste, and adjust the seasonings in the mixture.)

On a clean work surface form the mixture into a roll about 4 inches in diameter. Wrap it tightly in a double layer of good cheesecloth and tie the ends. Heat the stock in a fish poacher or other pan that will hold the meatloaf submerged in the stock and lower the loaf carefully into the liquid. Bring to a boil over medium heat, reduce the heat to low, and poach gently for 40 minutes. Turn off the heat and let the loaf cool in the stock for 1 hour. It should be quite warm, but not hot.

Remove the loaf from its liquid, let it drain for 10 minutes on a rack set over a plate, and add the liquid back to the stock. (Reserve the stock for another purpose, such as soup). Cut the meatloaf into thick slices and serve with Green Peppercorn Mustard alongside.

Variation:

This meatloaf also cooks up nicely in the oven. Pack the mixture into a 2-pound bread pan, and bake it in an oven preheated to 350 degrees until cooked through, about 50 minutes. Remove from the oven and it let rest, covered, for 10 to 20 minutes before slicing and serving with the mustard.

Classic Corned Beef and Cabbage

Serves 8

Here's the version of one of Ireland's signature dishes that I've been cooking for a couple of decades. It's almost impossible to mess up; just be sure to cook the corned beef long enough, so that it is very tender.

1 raw brisket of corned beef, about 4 pounds
2 teaspoons black peppercorns
3 whole small dried chiles or 1 teaspoon red pepper flakes
1 bay leaf
2 thyme sprigs
3 Italian parsley sprigs
6 large yellow onions, peeled and cut into sixths
3 medium carrots, peeled and cut into diagonal 2 ½-inch pieces
6 medium potatoes, scrubbed and cut into wedges
5 pounds cabbage, cored and cut into 3-inch wedges
½ cup heavy cream
3 tablespoons prepared horseradish
3 tablespoons minced fresh Italian parsley

Rinse the corned beef under cool tap water. Set it in a large pot, add the peppercorns, chiles, bay leaf, thyme sprigs, parsley sprigs, half the onions and enough water to come about 4 inches above the brisket. Bring to a full boil over high heat, reduce the heat to medium-low and use a large shallow spoon to skim off the foam and other impurities that rise to the surface.

Cover the pot, setting the lid slightly off center so that it is not a tight fit. Simmer 2 hours. Remove the lid and using a slotted spoon, remove and discard the onions and herb sprigs. Add the remaining onions and carrots and simmer, partially covered, for 30 minutes. Add the potatoes and simmer until they are almost tender, about 20 minutes. Add the cabbage, pressing it down into the liquid (it will rise back up but don't worry about it). Cover the pot and simmer 20 minutes, or until the cabbage is tender but not mushy.

Meanwhile, put the cream into a small bowl and stir in the horseradish and 1 tablespoon of the minced parsley. Taste, season with salt and pepper and set aside.

Use a large fork or tongs to transfer the brisket to a serving platter; cover it loosely with aluminum foil and let rest 15 minutes. Slice the corned beef and return it to the platter. Use a slotted spoon to transfer the vegetables from the pot to the platter; sprinkle the parsley over the vegetables. Serve immediately, with the horseradish cream on the side.

Variations:

- Instead of water, cook the corned beef in beer. I use a pilsner but have also enjoyed corned beef cooked in Guinness.
- Remove the corned beef before adding the potatoes to the cooking liquid and finish it in a 325-degree oven for 45 minutes; remove from the oven and let rest 15 minutes before slicing and serving.

A Salt & Pepper Cookbook

Standing Rib Roast (Prime Rib) with Carrots in Peppercorn Cream

Serves 6 to 8

For years, I've shared prime rib with friends on Christmas Day, sometimes at my home, sometimes at theirs. We always have several winter vegetables alongside, Brussels Sprout Leaves (page 303), mashed or smashed potatoes, and these voluptuous carrots.

4 to 5 garlic cloves, peeled and crushed
Kosher salt
4 teaspoons black peppercorns, lightly cracked
4 teaspoons white peppercorns, lightly cracked
1 tablespoon sweet paprika, preferably Spanish
1 tablespoon olive oil
3 or 4 rib beef roast, about 6 to 7 pounds
¾ cup heavy cream
2 ½ pounds carrots, preferably Nantes, peeled and cut into ¼-inch-thick diagonal slices
½ cup dry red wine
2 tablespoons butter
3 tablespoons Italian parsley, chopped
Black pepper in a mill

Preheat the oven to 450 degrees.

Put the garlic into a suribachi or large mortar, sprinkle with salt and use a wooden pestle to grind it into a paste. Add 2 teaspoons of black peppercorns and 2 teaspoons of white peppercorns; mix in the paprika and the olive oil.

Set the roast on a clean work surface and rub the mixture into it.

Set the roast, bone-side down, in a heavy shallow roasting pan, set in the oven and cook for 20 minutes. Reduce the heat to 325 degrees and cook 35 minutes more, without opening the oven.

Test the temperature of the meat with an instant-read meat thermometer; if it is between 115 and 120 degrees, remove from the oven. If not, cook 10 minutes more and test again. Continue until the temperature reaches 115 degrees for very rare and 120 for medium rare.

Remove from the oven and cover, loosely, with a sheet of aluminum foil. Let rest for 20 to 30 minutes, during which time the temperature of the meat will continue to rise.

While the meat cooks, put the cream into a small saucepan, add the remaining black and white peppercorns, set over medium heat and bring to a boil. Cover, remove from the heat and let steep.

When the roast comes out of the oven, put the carrots into a medium saucepan, add the wine and enough water to cover the carrots. Season with salt, set over medium heat and when the water boils, reduce the heat, cover and simmer the carrots gently until just tender, about 8 to 9 minutes.

Remove from the heat and drain. Strain the cream into the pan, return to a medium-low flame and heat through. Season with salt to taste, stir in the parsley and transfer to a warmed serving bowl. Grind black pepper over the carrots, cover and keep hot.

Carve the beef into ⅓- to ½-inch-thick slices, arrange on a platter, season with salt and pepper and serve immediately, with the carrots and other vegetables alongside.

Bratwurst with Sauerkraut & Apples

Serves 4 to 6

Bratwurst is one of the sausages that rely on white pepper for their flavor; in fact, a great deal of ground white pepper is sold to sausage makers both in the United States and western Europe to make these sausages. Alberger salt, too, plays an important role; because its crystals are hollow, it readily absorbs moisture and helps hold the ground meat and spices in emulsion so that the sausage has a uniform texture. Long slow cooking is crucial in this dish; undercooked, the ingredients fight with one another. Given enough time, the flavors mingle together, everything turns a sort of pale golden brown, and the dish becomes truly sensational. The ideal accompaniment would be crisp potato latkes and a crisp hard cider.

2 firm ripe apples, such as Granny Smith or Rome Beauty, peeled and cored
Half a lemon
2 pounds bratwurst (about 6 to 8 sausages)
3 cups fruity white wine (Viognier, Riesling, or Gewürztraminer)
2 ounces diced pancetta or bacon, optional
2 tablespoons butter
2 red onions, sliced
1 teaspoon freshly ground white pepper
½ teaspoon caraway seed
3 cups sauerkraut, homemade (page 277) or commercial
Dijon or other mustard

Cut the apples in lengthwise slices, cut the slices in half crosswise, and put them in a bowl of water. Squeeze the lemon into the water and set aside.

Place the bratwurst in a medium sauté pan, add 1 cup of the wine and 1 cup of water, set over medium-high heat, bring to a boil, cover, reduce the heat, and simmer for 10 minutes. Uncover, increase the heat to medium, and simmer until the liquid is nearly completely evaporated. Transfer the sausages to a plate and set it aside.

Add the pancetta or bacon to the pan and sauté for 7 or 8 minutes, until it is almost crisp. Use a slotted spoon to transfer to the plate with the sausage. Add the butter to the pan and, when it is melted, add the onions. Cook over low heat until the onions are very limp, fragrant, and sweet, about 20 minutes. Drain the sliced apples, add them to the onions, and sauté, stirring now and then, until they begin to brown, about 7 or 8 minutes.

Meanwhile, slice the sausages into ¼-inch rounds; slice the rounds in half.

Add the pepper and caraway seed to the onion mixture. Stir in the sauerkraut, add the sausages and pancetta, and stir in the remaining 2 cups wine. Bring to a boil over medium heat, reduce to a simmer, cover, and simmer very slowly for 45 minutes, until the juices have thickened and the flavors have mingled. Transfer to a serving dish and serve immediately, with mustard alongside.

A Contemporary Haggis with Neeps & Tatties

Serves 6 to 8

Shortly after I finished the manuscript for the first edition of this book, I traveled to Scotland, where I visited several distilleries and learned my way around single malt scotch, which I now thoroughly enjoy. On one of our last nights there, a special party celebrated 18th century Scottish poet Robert Burns, complete with bagpipes, kilts, a reading of "Ode to a Haggis," and the traditional parade of the haggis around the dining room before serving it. The haggis itself was quite good, contrary to its reputation, which comes, I think, from tinned versions and from early recipes, which include lamb lights, or lungs, an ingredient we typically do not eat today. This contemporary version honors the traditional one, yet with ingredients that are more readily available and acceptable to our modern sensibility. The signature ingredient, the spice that gives haggis its character, is black pepper. Don't skimp on it. Be sure to serve a single malt scotch alongside, even just a ceremonial jigger of it. I prefer Laphroaig, with its subtle taste of the briny sea and its smokiness from peat; it's like bacon in a glass.

3 or 4 lamb tongues
2 lamb kidneys
1 pound lamb shoulder
8 ounces lamb or calf liver
Kosher salt
1 teaspoon whole black peppercorns
6 ounces beef suet, chopped
2 yellow onions, minced

2 teaspoons fresh thyme leaves or 1 teaspoon dried thyme
1 ½ cups steel cut oats
Black pepper in a mill
Neeps and tatties (see Note on page 239)
Parsley sprigs

Put the tongues, kidneys, lamb shoulder, and liver into a medium pot, cover with water, add a tablespoon of salt and bring to a boil over high heat. Reduce the heat to low and simmer gently for 1 hour, skimming foam from the surface as it forms. Remove from the heat and let cool for 15 minutes.

Use tongs to transfer the tongues to a work surface; peel them, discarding the thin skin, which should come off easily.

Use a sharp chef's knife to mince all the meats. Set aside briefly. (Alternately, you can pass the cooked meats through a hand-cranked grinder if you prefer.)

Put a little of the beef suet into a heavy sauté pan—cast iron is ideal—and set over medium-high heat; when it is melted, add the onions, season with salt and sauté until soft and fragrant, about 10 minutes. Transfer to a small bowl, wipe the pan clean and return to medium heat. Add the oats and toast lightly, stirring all the while until they are fragrant and take on just a bit of color.

Stir in the minced meats, the suet, the thyme, and the cooked onions. Season very generously with black pepper; you want at least 2 teaspoons. Correct for salt.

Transfer the mixture to a glass or ceramic baking dish; a round Pyrex bowl or a 2-quart soufflé dish is ideal. Strain 3 cups of the poaching liquid over the mixture. Seal the dish tightly with a sheet of aluminum foil. Add a second sheet, crimping it tightly to the dish.

Fill the bottom of a steamer with water and add the steaming basket or rack. Set the haggis mixture in the basket or on the rack and set over high heat. When the water boils and steam begins to rise, reduce the heat to very low, cover and steam for 1 hour. Remove from the heat and carefully lift the foil to look inside. If the mixture seems dry, strain 1 cup of the poaching liquid over it. Crimp the foil, return to the heat and steam for an additional hour.

Remove from the heat and let rest 15 minutes before serving.

To serve, set a generous mound of haggis off center on individual plates, add neeps and tatties alongside, season everything with salt and pepper, garnish with parsley and serve.

NOTE

Neeps and tatties are, simply, yellow turnips or rutabagas and potatoes, boiled and mashed separately, seasoned with plenty of good butter, salt, and pepper and served side by side. To serve 6 to 8, you'll need about 1 ½ pounds of both turnips or rutabagas and potatoes. Feel free to have fun; the name encourages it, I think. You might use red turnips and blue potatoes or use just 1 pound of turnips mashed with ½ pound of carrots, along with the potatoes, of course.

Salt Blocks, Bowls & Plates

In the winter of 2007, my dear friends and colleagues, Ridgely Evers and Colleen McGlynn of DaVero, producers of some of the finest olive oils in the world and also of outstanding Italian-varietal wines in the heart of Sonoma County, spent a week sailing in the Caribbean. It was there that Colleen, an accomplished chef, came across a beautiful slab of marbled Himalayan salt. She made ceviche on it, was impressed with how it absorbed just the right amount of salt and loved, too, the dazzling presentation.

She was sold and began offering salt plates of every shape and size, from tiny sushi sticks, which can be chilled in the freezer just before topping with individual portions of, say, hamachi or maguro, to large platters big enough to hold a whole salmon in the oven, at Supper, a high-end take-out café she and Ridgely operated for a time in downtown Healdsburg.

Now, salt plates are everywhere, or nearly so, where passions for artisan foods and unique experiences run high. At The Meadow, which has two locations in Portland, Oregon, one in Manhattan, and an online store, you'll find shot glasses, goblets, single-serving bowls, mixing bowls, plates and platters, tiny cubes, large rounds, and even little salt cellars. It's a salt lover's wonderland.

These salt blocks are versatile, durable, and tested for both high- and low-temperature tolerance. They are beautiful, fun, and dramatic to use and, in the end, ephemeral. They will eventually disappear, as the salt is transferred to the foods that they hold and to the water that must be used to clean them between each use. When they crack, which is inevitable, they can be broken into smaller pieces to use as rock salt and grated for cooking salt.

They are also expensive. A set of four goblets runs nearly sixty dollars and a single small bowl is around twenty dollars, enough for eight guests inching above one hundred and fifty dollars, a pretty price to pay for something that vanishes a bit with each use.

A Salt & Pepper Cookbook

What salt blocks are not is essential. If money is not a concern and you love dazzling your dinner guests, have at it. But if you live on a tight budget as so many of us do, don't feel that your cooking will be compromised because you can't make your aioli in a salt bowl or serve your Bloody Marys in salt goblets. If you're curious, start with a medium-sized block and experiment with different preparations.

I have just a single salt block in my kitchen but I confess that I keep a salt lamp on my desk. Why? Because it is beautiful, because its smooth round shape evokes the moon, and because it inspires me.

A warning is in order. You must protect your salt vessels and blocks from humidity. If you live in a particularly moist climate, be especially careful to dry them thoroughly between uses and to store them in the driest area possible.

Salt can be used to encase foods, creating a tiny oven that holds in aromas and prevents juices from evaporating.

Cooking and Curing In & On Salt

We think of salt primarily as a seasoning but it is also used for its ability to denature protein—that's what heat does, too, and acid; casually, it's called cooking—and to seal in flavor. In the first section of this chapter, I employ salt in various cooking techniques, with several types of salt crusts and, more simply, with a salt block, that both seasons and transforms, just a bit, the foods that are set upon it.

In the second section, I explore salt's transformative qualities, its ability to cure, to preserve, and to encourage fermentation, techniques that have been used for centuries and that are remarkably easy and reliable to do at home. Not that long ago, most home cooks had pantries full of preserved foods. It's only been in the last century that the practice has all but disappeared in most of America. It is now enjoying an exciting renaissance, as so many of us long to return to natural foods grown close to home in their own natural time and preserved in simple ways at their peak of flavor.

Cooked

Carpaccio Beet
Simple Halibut Ceviche
Shrimp Roasted on Rock Salt
Snapper Fillets Baked in a Salt Crust
Trout Baked in a Salt Crust
Rockfish Baked in Salt
Lamb Loin Baked in a Salt Crust
Beef Tenderloin in a Salt Crust
Whole Salmon Baked in a Salt Crust
Seasoned Game Hens Roasted in Salt

Cured

Salted Anchovies
Smoky Salmon Gravlax
Halibut Gravlax
Bread & Butter Pickles
Patrick's Serrano Pickles
Asian Salt Eggs
Homemade Sauerkraut
Salted Mustard Greens
Preserved Lemons
Yogurt Cheese
Brined Pork Tenderloin
Corned Beef & Pastrami (Sidebar)
Homemade Bacon

Beet Carpaccio

Serves 2 to 4

For a long time, carpaccio was synonymous with very thinly sliced beef tenderloin drizzled with good Italian olive oil and seasoned with salt, pepper, and curls of Parmigiano-Reggiano. In recent years, the term is used more as a description of technique and may be anything that is sliced very thinly: beets, zucchini, tomatoes, watermelon, salmon, lamb, goat, venison, and, of course, beef. It can be crowned with almost any condiment or combination of herbs. Typically served on a plate, carpaccio is also delicious served on a salt block.

1 medium golden beet, trimmed and peeled

1 medium Chioggia beet, trimmed and peeled

Olive oil

1 small to medium salt block, about 7 inches square or thereabouts

Blood Orange Olive Oil or best-quality extra virgin olive oil

Zest of 1 orange

Flake salt

Black pepper in a mill

Preheat the oven to 375 degrees. Put the beets into a small baking dish, drizzle with just enough olive oil to coat them and cook until tender, about 45 minutes. Remove from the oven and let cool until easy to handle.

Chill the salt block. Using a very thin very sharp knife, cut the beets into the thinnest rounds possible, no more than ⅟₁₆ of an inch.

Arrange the beets on the salt block, making two rows, one of slightly overlapping golden beets and one of slightly overlapping Chioggia beets. Chill until ready to serve.

Remove from the refrigerator and drizzle with a bit of olive oil. Scatter the zest on top, season with salt and pepper and serve right away, as a starter course or side dish.

Simple Halibut Ceviche

Serves 2, easily increased

There are so many versions of ceviche, from a simple bit of raw fish sliced paper thin and seasoned with lime juice and salt to the classic ceviches of Peru, which can be a meal in themselves and feature many ingredients, fish cooked in acid and salt, chiles, herbs, sweet potatoes, corn, garlic, onions, and more. All are delicious, good enough to make you wonder why fish is ever subjected to heat.

In this version, both acid and salt are used in the cooking and seasoning. It is lovely on a hot summer night, with sparkling wine, a big green salad, and fresh seasonal fruit to complete the meal.

4 ounces very fresh Pacific halibut fillet, sliced very very thin
Juice of ½ lime
1 small to medium salt block, about 7 inches square or thereabouts, chilled
¼ lime, cut into tiny thin wedges
½ serrano, minced
Flake salt
Black pepper in a mill

Set the halibut on a plate, pour the juice over it and turn the slices of fish over once or twice so that they are coated in the lime juice. Let rest for 5 minutes.

Transfer the fish to the salt block and refrigerate until ready to serve, but no longer than 30 minutes.

Remove from the refrigerator and set a tiny lime wedge on each piece of fish. Scatter serranos over everything, season with salt and several turns of black pepper and serve right away.

After enjoying, wipe the salt block clean and store it in a cool, dry cupboard.

Shrimp Roasted on Rock Salt with Spicy Maple Butter

Serves 2 to 4

Shrimp roasted on a bed of hot salt are succulent, tender, fragrant, and not at all salty; the salt serves to transfer heat efficiently, not to flavor the shrimp. When you add spices or herbs, they will perfume the shrimp with their aromas, which you can echo in an accompanying sauce. This is my favorite version, with a sauce that is finger-lickin' delicious but you should feel free to add whatever you prefer, crushed red pepper flakes, cinnamon sticks, star anise, dried thyme leaves, or dried lavender, for example, to the salt and to the butter.

2 pounds rock salt of choice
8 cardamom pods, cracked open
1 tablespoon white peppercorns, cracked
1 tablespoons black peppercorns, cracked

2 teaspoons whole cloves
12 large shrimp, preferably wild Gulf shrimp, in their shells, heads on
Spicy Maple Butter (page 356)
Lemon wedges, for garnish

Put the salt into a tempered clay pot, preferably one that is about 2 ½ to 3 inches deep and long enough to hold the shrimp in a single layer. Add the cardamom pods, peppercorns, and cloves, toss and spread evenly.

Set in the oven, turn the temperature to 400 degrees and heat for 45 minutes, until the salt is very, very hot.

Working quickly, set the shrimp on top of the hot salt, close the oven and cook for 3 minutes. Use tongs to turn over the shrimp; cook for 3 to 4 minutes more, until they have turned opaque.

While the salt heats, make the sauce, if you have not already done so.

Turn off the oven. Use tongs to transfer the shrimp to a serving platter, drizzle with some of the sauce, garnish with lemon wedges and serve right away, with the remaining sauce alongside.

Snapper Fillets Baked in a Salt Crust

Serves 4 to 6

You can bake any fish—filleted or whole—in a crust of salt without wrapping it. However, it is easier, especially for the home cook who may use this technique only occasionally, to get the hang of is if the fish is wrapped in something that will shield it from direct contact with the salt. Although the fish itself does not become salty (that is, the salt does not penetrate or flavor the flesh), a substantial quantity can cling to the outside of the fish and it can be tedious to brush it off. I suggest protecting both fillets and whole fish from the salt by setting them on top of fresh or dried corn husks, fresh grape leaves, dried (but not processed) seaweed, or sprigs of fresh herbs and then adding a layer on top of the fish. This protects the fish, makes a beautiful presentation and, with herbs, adds layers of flavor and aroma. In this recipe, the salt is combined with egg whites and water, which makes a harder crust than one with just salt.

6 snapper fillets, about 6 ounces each
2 teaspoons crushed black peppercorns
1 teaspoon kosher salt
1 teaspoon ground cumin
½ teaspoon cayenne pepper
3 or 4 pounds rock salt or kosher salt
2 egg whites
6 to 8 large corn husks or 12 large fresh grape leaves
1 bunch cilantro
2 lemons, cut into wedges

Preheat the oven to 400 degrees. Set the snapper fillets on a work surface. In a small bowl, mix together the pepper, kosher salt, cumin, and cayenne; sprinkle all over the fillets. In a large bowl, combine the salt, egg whites, and ½ cup water, mixing to form a loose, sticky paste. Spread a layer of this salt paste about ¾ inch thick on the surface of a large baking sheet or other ovenproof container. Put a layer of corn husks or grape leaves down the center and set the fillets on top of the leaves; cover the fillets with the remaining grapes leaves and add the remaining salt paste on top, completely burying the fish.

Set on the middle rack of the oven, bake for about 20 minutes, testing with a thermometer and removing the fish from the oven when the temperature is about 130 degrees. Let rest out of the oven for 10 minutes. Carefully crack open the salt and extract the fish fillets.

Spread the cilantro on a serving platter, set the fillets on top, garnish with lemon wedges and serve right away.

Trout Baked in a Salt Crust

Serves 2

    I love wild trout and when I can get it, that is what I use in this dish. Branzino is a good substitute. Before you buy any seafood, I recommend consulting Seafood Watch (seafoodwatch.org) for the best choices and good alternatives. The nonprofit organization, a project of the Monterey Bay Aquarium, monitors fish populations, fishing techniques, and farming practices around the world and posts their recommendations annually. I find them a trustworthy source for up-to-date environmental information.

    In this dish, rock salt is a good choice because it makes the best-looking crust, but you can use whatever is least expensive—flavor, finesse, texture, or other subtle qualities of specific salts are irrelevant when salt is used in this way. Most recipes for salt crust call for the food—a loin of beef, a whole chicken, a fillet of salmon, a peach—to be placed directly on the salt. In a few recipes, the food is encased first in herbs, then in parchment, and lastly in foil before being tucked into its nest of salt. I prefer a middle ground, with something between the salt and the food itself that will both protect it and contribute aroma and flavor. If you happen to grow French lavender, you can use clippings of the greens (not the flowers) in place of the parsley. It will impart a subtle yet appealing flavor and aroma to the fish. You can use fresh or dried corned husks, fresh grape leaves, or other fresh herbs.

2 trout, about 1 to 1 ¼ pounds each, cleaned but not boned
4 slices fresh ginger
3 pounds rock salt
2 egg whites
2 lemons, thinly sliced
1 bunch flat-leaf parsley, with stems
1 lemon, cut into wedges
½ cup Spicy Salt Egg Sauce (page 355), optional

Rinse the trout under cool water and dry on a tea towel. Place two slices of ginger inside the cavity of both trout.

Preheat the oven to 375 degrees. In a medium bowl, combine the salt, egg white, and ⅓ cup water. The mixture should be slightly sticky and without

     A Salt & Pepper Cookbook

lumps. Place about a third of the salt on a baking dish or baking sheet in a ¾-inch-thick rectangle just slightly larger than the 2 trout. Leaving a 1-inch margin, cover the surface of the salt with half of the lemon slices and set the trout on top. Cover the trout with the remaining lemon slices and cover any remaining exposed skin with some of the parsley.

Add parsley sprigs between the trout, tucking them in as tightly as possible. Pack the remaining salt over the trout, enclosing it completely.

Set on the middle rack of the oven and bake for 30 minutes. Remove the pan from the oven and let it sit for 5 minutes before breaking open the salt, which will have hardened, and removing the trout. Use a dry pastry brush to brush off any salt that may have stuck to the trout.

Set the trout on a serving platter, garnish with lemon wedges, and serve immediately, with the Spicy Salt Egg Sauce, if using, on the side.

Rockfish Baked in Salt

Serves 3 to 4

Years ago I was visiting a new friend and when we were hungry, she pulled a few sheets of nori from the cupboard, tore them in half, and toasted them over the flame of a gas burner. They turned black and crisp almost instantly. Next, she poured soy sauce into a small bowl and squeezed the juice of a lime into it. We dipped the nori into the sauce and I thought it was one of the most delicious, refreshing things I'd ever tasted. The memory of that long-ago pleasure inspired this recipe. The nori shields the fish from direct contact with the salt, making its removal all the easier, the sauce couldn't be simpler, and it is perfect with the tender fish. The salt crust in this version is made of salt only, without the water and egg white that are used in the other recipes in this book. This method is simple and direct, and you get excellent results, but the addition of water and egg white does make a more solid crust. Feel free to use that technique if you prefer.

1 rockfish (also known as Pacific snapper, rock cod, and ocean perch), dressed
Kosher salt
Black pepper in a mill
3 slices fresh ginger
4 sheets nori
4 pounds rock salt
½ cup dark soy sauce
Juice of 1 lime
1 tablespoon grated fresh ginger
1 bunch (about 10) scallions, trimmed and sliced into thin rounds

Preheat the oven to 400 degrees. Rinse the fish in cool water and pat it dry with a tea towel.

Season the inside of the fish with salt and pepper and tuck in the ginger.

Spread a layer of salt about ¾-inch thick on the surface of a large baking sheet or other ovenproof container large enough to hold the fish. Wrap the fish entirely in the nori, set it on top of the salt, and then use the remaining salt to bury the fish completely.

Bake for about 30 to 40 minutes, until the fish reaches an internal temperature of about 130 degrees (use an instant-read thermometer and poke it through the salt crust into the fish). Remove the pan from the oven and let it rest for 5 to 10 minutes, during which time the fish will continue to cook.

Meanwhile, mix together the soy sauce, lime juice, ginger, and about a tablespoon of the scallions. Carefully remove the fish from its salt bed, set it on a work surface, and remove the nori wrapper. Set the fish on a serving platter, scatter the remaining scallions over the surface, and serve immediately, with the sauce alongside.

Lamb Loin Baked in a Salt Crust

Serves 3 to 4

Lamb is often described as "spring lamb," a designation that has nothing to do with the time of year. Rather, it describes the age of the lamb at harvest. Spring lamb is from animals that are between four and six months old; because lambs were traditionally born in December and January, they received the seasonal designation. Today, the breeding of flocks is controlled and spring lamb is available nearly year round. As delicate and flavorful as spring lamb is, there's a prejudice against the meat that lingers from decades past when most lamb came from yearlings, older animals with a strong, gamy smell that many people find offensive, an aroma absent from young lambs.

Loins from spring lamb can be fairly small; you may need to place two side by side to make this dish.

1 ½ cups kosher salt
½ cup minced fresh herbs (rosemary, oregano, and flat-leaf parsley)
2 tablespoons freshly ground black pepper
2 ½ cups all-purpose flour
2 egg whites
2 tablespoons olive oil
1 lamb loin, boned, about 1 ¼ pounds
Fleur de sel or sel gris
Black pepper in a mill
Fresh herb sprigs, for garnish

Place the kosher salt, ¼ cup of the minced herbs, the black pepper, and flour in the bowl of a heavy-duty mixer fitted with a paddle. Mix briefly. In a small bowl, whisk together the egg whites and ½ cup plus 2 tablespoons water until slightly foamy. With the mixer operating at low speed, slowly pour the egg and water into the salt mixture. When the water is fully incorporated, increase the speed to high and knead until a firm but moist dough is formed, about 3 or 4 minutes. Turn out onto a very lightly floured surface, form into a ball, and cover with plastic wrap. Let the dough rest at room temperature for at least 4 hours and up to 24 hours.

Preheat the oven to 400 degrees. Heat the olive oil in a heavy skillet over medium heat and sear the lamb loin. Transfer to a plate and set aside. Roll out the salt dough on a lightly floured surface to form a rectangle approximately 8 by 11 inches (adjust the size according to the lamb). Set the loin in the center of the dough, sprinkle the remaining ¼ cup of herbs over the it, turning it to spread the herbs evenly over the surface of the meat, and wrap the dough tightly around the loin, sealing the edges by pinching the dough together. Set on an ungreased baking sheet and bake for 20 minutes for rare lamb, 25 for medium rare.

Remove from the oven and let the lamb rest for 20 minutes before removing the salt crust (keep in mind that the meat will continue to cook in its crust of salt). Slice off one end of the crust, pull the lamb out, and cut it into thick slices. Arrange on a platter, season lightly with salt and several turns of black pepper, garnish with fresh herb sprigs, and serve immediately.

Beef Tenderloin in a Salt Crust

Serves 3 to 4

A mixture of salt, flour, and egg whites forms a seamless crust that keeps this beef tenderloin tender and juicy. The salt dough can be made with or without the herbs, which contribute subtle flavors and aromas. Once removed from the oven, the meat continues to cook inside its crusty shell and will remain hot for as long as an hour. If it is to sit for a time before you serve it, be sure to remove it from the oven before it reaches the desired temperature, or the meat will be overcooked by the time you put it on the table.

1 ½ cups kosher salt
½ cup minced fresh herbs (rosemary, oregano, thyme, and Italian parsley)
2 tablespoons freshly ground black pepper
2 ½ cups all-purpose flour
2 egg whites
2 tablespoons olive oil
1 ½ pounds beef tenderloin, trimmed
Fleur de sel or other finishing salt
Fresh herb sprigs, for garnish

Put the kosher salt, ¼ cup of the herbs, the black pepper, and flour in the bowl of a heavy-duty mixer fitted with a paddle. Mix briefly. In a small bowl, whisk together the egg whites and ½ cup plus 2 tablespoons water until slightly foamy. With the mixer operating at low speed, slowly pour the egg and water into the salt mixture. When the water is fully incorporated, increase the speed to high and knead until a firm but moist dough is formed, about 3 or 4 minutes. Turn out onto a very lightly floured surface, form into a ball, and cover with plastic wrap. Let the dough rest at room temperature for at least 4 hours and up to 24 hours.

Preheat the oven to 400 degrees. Heat the olive oil in a heavy skillet over medium heat and sear the tenderloin on all sides. Transfer it to a plate and set it aside. Roll out the salt dough on a lightly floured surface to form a rectangle approximately 8 by 11 inches (adjust the size according to the size of the beef). Set the beef in the center of the dough, sprinkle the remaining

¼ cup of herbs over the beef, turning it to spread the herbs evenly over the surface of the meat, and wrap the dough tightly around the tenderloin, sealing the edges by pinching the dough together. Set on an ungreased baking sheet and bake for 20 minutes for rare beef, 25 for medium rare.

Remove the tenderloin from the oven and let rest for at least 30 minutes and up to 1 hour before removing the salt crust. Slice off one end of the crust, pull the beef out, and cut it into thick slices. Arrange on a platter, sprinkle lightly with fleur de sel, garnish with fresh herb sprigs, and serve immediately.

Serving Suggestions: This tenderloin goes beautifully with Risotto Pepato, page 176. In the spring, serve Roasted Asparagus, page 288, too, and at other times, consider adding Glazed Carrots, page 291, or Black Pepper Zucchini, page 293, alongside.

Variation:

To cook a whole chicken in a salt crust, add the minced zest of 1 lemon to the dough and let the dough rest overnight. Rinse a 3-to 4-pound chicken under cool running water and dry on a tea towel. After seasoning the cavity with salt and pepper, fill it with half a yellow onion cut into wedges, half a lemon cut into wedges, and several sprigs of flat-leaf parsley. Season the outside of the chicken with olive oil and freshly ground black pepper. Roll the dough out so that it is large enough to fit loosely around the chicken. Set the chicken, breast-side up, in the center. Fold the dough up and over the chicken and seal it—it should be loose, not fit the chicken too tightly. Set the bird on a rack in a roasting pan and bake at 400 degrees until the chicken reaches an internal temperature of 140 degrees in the thickest part of the thigh. Remove the chicken from the oven and let it rest between 15 and 30 minutes, during which time it will continue to cook. Remove the crust and serve immediately.

Whole Salmon Baked in a Salt Crust

Serves 6 to 8

I used to bury salmon in a pit in the ground, which produces excellent results but is quite time consuming; it also requires a place in which to dig the pit, which means apartment dwellers are out of luck. Anyone, however, can encase a salmon, or other fish, in salt, which serves many of the same functions as the underground method; it keeps the salmon juicy, traps aromas, and provides the pleasure of unearthing a treasure. For this recipe, you will need a large pan; what is called a hotel pan (an insert for a chafing dish that is 2 inches deep) works well, but not everyone has one sitting around the house. Use the longest pan you have, and if the tail of the salmon hangs over, just wrap it in aluminum foil so that it doesn't burn.

8 to 10 pounds salt
3 egg whites
18 large fresh grape leaves
One 5- to 6-pound fresh wild Pacific King salmon
Kosher salt
Black pepper in a mill
12 cilantro sprigs
2 stalks lemongrass, cut into ½-inch pieces
8 garlic cloves, lightly crushed
1 small red onion, cut in wedges
3 serranos, split
2 lemons, cut into wedges

Preheat the oven to 400 degrees. In a large bowl, mix together the salt, egg whites, and ¾ cup water. Spread a layer of salt about ¾-inch thick on the surface of a large baking sheet or other ovenproof container large enough to hold the fish.

Put a layer of grape leaves down the center of the salt.

Season the cavity of the salmon with salt and tuck in the cilantro sprigs, lemongrass, garlic, onion, and serranos. Set the salmon on top of the grape leaves, and salt and cover it with the remaining grape leaves. Be sure to tuck the leaves around the curves of the fish. Top with the remaining salt, covering the salmon completely.

Bake for about 50 to 60 minutes, until the fish reaches an internal temperature of about 130 degrees (use an instant-read thermometer and poke it through the salt crust into the fish). Remove the pan from the oven and let rest for about 15 minutes, during which time the fish will continue to cook.

Carefully break the salt crust and remove the salmon from the salt. Set it on a work surface, remove the grape leaves, and set the salmon on a serving platter. Serve immediately, with lemon wedges alongside.

Variation:

To serve chilled, remove the salmon from the salt and remove the aromatics from the salmon's cavity. Set it on a platter, cover it, and chill it thoroughly. Serve it with lemon wedges and Green Peppercorn Mayonnaise alongside.

Seasoned & Stuffed Game Hens Roasted in Salt

Serves 4

    These succulent little hens are so incredibly delicious you can serve them with nothing more than wedges of lime, corn tortillas, and rice, or you can serve a more complex condiment to complement them. Grilled Corn with Pepper Butter (page 297) is perfect alongside, and Pineapple with Black Pepper (page 101) makes a refreshing starter. They are also outstanding when stuffed with polenta; a perfect time to make them is when you have leftover polenta (see page 189). Just cut polenta into cubes, toss it with some chopped Italian parsley or cilantro, and pack it into the central cavity, not filling it too full, as the polenta will expand.

1 ¼ cups, approximately, leftover polenta (page 189)
Kosher salt
Black pepper in a mill
4 Cornish game hens
2 tablespoons olive oil
1 tablespoon smoked paprika or other paprika of choice
1 package dried corn husks, soaked in hot water for 2 or 3 hours and
    drained
5 to 6 pounds rock salt or kosher salt
2 egg whites

Preheat the oven to 375 degrees.

If the polenta is quite firm, cut it into ½-inch cubes and season lightly with salt and pepper. Set aside. Rinse the game hens under cool water and let dry on tea towels.

In a small bowl, mix together the olive oil, paprika, a teaspoon of freshly ground black pepper, and 2 teaspoons of salt and rub the mixture into the skin of each hen. Set several corn husks on your work surface, overlapping them. Set a game hen in the center and fold the husks over the hen, tying each end tightly either with thin strips of husk or with string. Wrap all of the hens and set them aside.

In a large bowl, combine the rock salt, egg whites, and ½ cup water until it forms a sticky mixture. Spread about a third of the mixture on the bottom of a pan large enough to hold the game hens with about 2 inches of space between them. Use the remaining salt to completely cover the hens.

Bake for about 30 minutes (10 minutes longer if you have stuffed the birds with polenta), until the internal temperature of the hens reaches about 140 to 145 degrees (poke through the salt crust directly into one of the hens). Remove the pan from the oven and let it rest for at least 15 minutes, during which time the hens will continue to cook, before breaking the salt crust and extracting the hens. Cut the ties and serve the hens immediately on the opened corn husks.

After removing from the oven, the salt encasing is hard and must be cracked open.

Use a heavy spoon or other sturdy implement to lift out the food that has been encase in salt.

Transfer a single portion to a plate and then carefully brush off salt that sticks to the wrapping

After the salt is brushed away, open the encasing--here, dried corn husks--and breath in the beautiful aromas.

Cured

About Brine-Cured Meats

Salted Anchovies

If you happen upon some very fresh anchovies—an unlikely occurrence in the United States except in certain ethnic and specialty seafood markets—you can preserve them in salt. First, remove the heads, gut them, rinse them in cool water, and set them on a tea towel to drain. Place a layer of flake salt ¼-inch deep in the bottom of a nonporous, nonreactive container (a large, wide-mouthed canning jar works well), followed by a single layer of anchovies. Sprinkle generously with salt; repeat until you've layered all of the fish; end with a final quarter-inch layer of salt. Store in a cool, dark pantry. The anchovies will be ready in ten to fourteen days and will keep for several months.

Smoky Salmon Gravlax

Serves 8 to 10

Gravlax is traditionally made with two large fillets of salmon coated in a salt mixture and then pressed together. Two full-length fillets make a lot of gravlax and I experimented to see how effective the process would be using a single piece. It works perfectly and is a great way to produce a smaller quantity. In this version, I use both smoked salt and a peaty single malt Scotch to add a smoky flourish to the salmon; for a more conventional version, consult the Variation at the end of this recipe.

Many traditional recipes for gravlax also call for fresh dill sprigs, which I do not use during the curing process, though sometimes I have used fennel fronds. I find dill, which is quite strongly flavored, eclipses salmon instead of enhancing it. If you like it, serve some minced fresh dill alongside when serving the gravlax.

1 wild Pacific King salmon fillet, about 1 ¾ to 2 pounds, skin on, scaled
3 tablespoons Laphroaig Scotch whiskey or other Islay Scotch
6 tablespoons kosher salt or other flake salt
2 tablespoons smoked salt
2 tablespoons sugar
2 tablespoon coarsely crushed black peppercorns

Use needle-nose pliers to gently pull out the salmon's pin bones, using your fingers to locate the bones.

Put the salmon, skin-side down, in a glass container in which it can lie flat and pour the whiskey over it.

Put the the salts, sugar, and peppercorns into a small bowl or other container, mix and spread over the salmon, adding the most where the filet is thickest; use all of the mixture.

Cover the salmon with a piece of plastic wrap and place a heavy weight on top. Cover the entire dish with plastic wrap and set in the refrigerator, After 1 day, remove the salmon from the refrigerator and unwrap it. Spoon the juices over the salmon and then turn it skin-side up. As before, cover it, weight it, cover the dish, and return to the refrigerator. Repeat this daily for up to 4 days (3 should suffice), by which time the flesh of the salmon will glisten. Remove the fish from the brine, wrap it tightly in clean plastic wrap, and refrigerate.

Kept wrapped tightly in clean wrap, the gravlax will keep for 8 to 10 days.

To serve the salmon, cut several very thin slices, cutting down to, but not through, the skin. Arrange them on a plate and serve with condiments, suggested below, alongside.

Variation:

To make a more conventional gravlax, use all kosher salt and replace the whiskey with vodka or gin. Add 3 or 4 crushed juniper berries to the black peppercorns.

Halibut Gravlax

Serves 16 to 20

Halibut cured in the same way as salmon produces delicate, delicious gravlax, with a brighter flavor than salmon. Most supermarkets sell halibut steaks rather than fillets, but if you have a specialty fish market or wholesaler near you, you should be able to order fillets (though you might have to purchase an entire fish).

2 halibut fillets, about 1 ¾ pounds each, skin on, scaled
2 tablespoons vodka
¾ cup kosher salt
4 tablespoons sugar
1 tablespoon toasted, crushed, and sifted Sichuan

peppercorns (see Note)
1 tablespoon coarsely crushed black peppercorns
1 red onion, thinly sliced
Dijon mustard
Black pepper crostini or crackers

Use needle-nose pliers to pull out any pin bones. Place one of the halibut fillets, skin-side down, in a glass or stainless-steel container in which it can lie flat and pour the vodka over it. Set the other fillet on a clean work surface. Combine the salt, sugar, and peppercorns. Spread half the mixture over each fillet, adding slightly more where the fillet is thickest, and rubbing slightly to ensure that the mixture sticks to the fish. Invert the second fillet on top of the fillet in the dish. Cover the halibut with a piece of plastic wrap and place a heavy weight on top. Cover the entire dish with plastic wrap and place in the refrigerator. After 1 day, remove the fish from the refrigerator and unwrap it. Spoon the briny juices over the halibut and then reverse the fillets, placing the one that had been on top on the bottom. Always keep the skin-side facing out, with the inner part of the fillets touching each other. As before, cover and weight the fish, cover the dish, and return to the refrigerator.

Repeat this daily for up to 4 days (3 days will probably suffice), by which time the flesh of the halibut will glisten. Remove the fish from the brine, wrap it tightly in clean plastic wrap, and refrigerate. Kept wrapped tightly in clean wrap, the gravlax will keep for 8 to 10 days.

A Salt & Pepper Cookbook

To serve the halibut, cut several thin slices, cutting down to, but not through, the skin. Arrange them on a plate and serve with onion, mustard, and crostini or crackers alongside.

NOTE

Unlike true peppercorns, Sichuan peppercorns, sometimes called fagara, have hard, tasteless seeds in their centers that should be sifted out and discarded after toasting and grinding.

Variation:

Serve with Artichoke & Olive Tapenade, page 353, or Green Peppercorn Mustard, page 358, alongside instead of the other condiments.

Bread & Butter Pickles

Makes 6 to 8 pints

Taste in pickles is remarkably personal. Who knows why you prefer crunchy sweet pickles seasoned with garlic while I crave salty kosher dills? As a kid, one of my favorite snacks was the big dill pickles I could buy for fifteen cents from the butcher counter at the little market around the corner. It wasn't until I was an adult that I developed a taste for these salty-sweet bread and butter pickles, which are one of the easier pickles to make at home.

1 gallon (about 5 pounds) pickling cucumbers, sliced ¼ inch thick
4 cups (about 4) yellow onions, peeled and cut into ⅜-inch lengthwise slices
1 ½ cups kosher salt
1 quart apple cider vinegar

3 thin slices fresh ginger
3 cups granulated sugar
1 tablespoon black peppercorns
1 tablespoon celery seeds
1 tablespoon white mustard seeds
1 tablespoon ground turmeric

Place the cucumbers and onions in a large bowl. Sprinkle 1 cup of the salt over them, cover with plastic, and add a weight, such as a heavy ceramic plate. Refrigerate overnight. Drain, rinse in cool water, and drain again, letting the vegetables sit in a strainer or colander until no more liquid leaches out.

In a large pot, combine the remaining ½ cup salt, vinegar, ginger, sugar, peppercorns, celery seed, mustard seed, and turmeric, set over medium heat, and stir continuously until the salt and sugar are dissolved. Reduce the heat to low, add the cucumbers and onions a large handful at a time, stir very gently and quickly, heat through, and remove from the heat. Cool slightly and pack into sterilized pint glass jars. Cover with the cooking liquid, seal the jars, and process in a boiling-water bath for 15 minutes (see Note). Remove the jars from the bath, let them cool, and check the seals. Let the pickles mature in a cool, dark pantry for a couple of weeks before using. If the seal is not secure, if the lid bulges, or if the pickles have an off smell, discard the pickles.

NOTE

When canning, be sure that jars are sterilized in boiling water immediately before they are used. Fill them within half an inch of the top of the jar. Use new seals dipped into boiling water and clean rings to close the jars. To process canned foods in a water bath, the water should come at least two inches above the jars. It is best to use a canning pot fitted with a rack; if jars come in contact with the bottom of the pot, they can crack. A rack will also keep the jars upright. Use tongs to remove jars from the pot, and set them on a wooden cutting board or a tea towel to cool. After they are cooled, check the seals by tapping on the top; if it sounds dull, the jar probably needs to be reprocessed or else stored in the refrigerator and used within a few days. Never taste canned foods with bulging lids; discard them immediately.

Patrick's Serrano Pickles

Makes about 1 quart

This recipe came to me from Mexico, by way of my dear friend Patrick Bouquet, who got it from his ex-wife's aunt, who used slightly moist sea salt from Baja California to make her pickles. The crucial factor is that she is familiar with the size and feel of the salt and measures only with her fingers. "Three generous pinches" is how Patrick described it as he watched her make the spicy pickles. Korean sea salt, which is very moist and with about the same size crystals as sel gris, is a good alternative. It's easy to find in Asian markets and sells for about a dollar for two pounds.

If you don't have serranos, use jalapeños or any similar chile. Patrick's aunt used jalapeños but I eventually switched to serranos, as I prefer their flavor. To use jalapeños, you'll need about 18.

30 fresh green serrano chiles, the
    longest ones you can find
3 tablespoons mild olive oil
4 to 5 large garlic cloves
2 teaspoons coarse, moist sea salt,
    such as Celtic gray or Korean
1 cup apple cider vinegar

2 tablespoons dried oregano
1 tablespoon dried thyme
2 tablespoons dried marjoram
3 to 4 bay leaves
3 to 4 carrots, peeled and sliced on
    bias
1 sliced yellow onion

Using a sharp paring knife, make a lengthwise slit from end to end in each serrano.

Heat the oil in a large heavy frying pan set over medium-low heat. Add the peppers and garlic, and sweat them, turning frequently, until the peppers begin to soften slightly and release their aroma, about 15 minutes. (Do not let the garlic burn.) Add the salt and vinegar. Agitate the pan and, when the salt is dissolved, add 1 cup of water, the oregano, thyme, marjoram, and bay leaves, and bring to a boil. Reduce the heat and simmer the until the serranos turn greenish yellow. Add the carrots and onion, remove the pan from the heat, and allow the mixture to cool to room temperature. Pack into a glass quart jar, pour all of the liquid over the vegetables. and seal the jar. These pickles will keep, refrigerated, for several months.

                                A Salt & Pepper Cookbook

Asian Salt Eggs

Makes 1 dozen

Salt eggs, made with duck or chicken eggs, are found throughout China and South-east Asia. In Malaysia, I found baskets of what looked like black golf balls; they were salted chicken eggs with a thick coat of black ash to preserve them longer. Most of the ash is easily brushed off, and then the eggs can be soaked in cool water for a few minutes; wipe the wet eggs clean with a tea towel. The eggs are then hard-cooked, shelled, and served as an accompaniment to a variety of dishes. They may also be topped with a dressing of garlic, shallots, serranos, and lime juice, and served as a simple salad. They can also be fried. Although we no longer need this technique to preserve eggs, it produces a pleasantly intense flavor that is not duplicated by merely adding salt to fresh eggs.

1 ¼ cups kosher salt or 1 cup pickling salt (see glossary, page 404)
4 quarts water
12 eggs

Combine the salt and water in a large pot, bring to a boil, remove from the heat, and let cool completely. Place the eggs in a large earthenware or glass jar, pour the brine over the eggs, cover the jar with a tight-fitting lid, and store in a cool, dark cupboard or pantry for at least 4 weeks and up to 3 months.

To use the eggs, boil them in tap water for about 12 minutes, until hard-cooked. Remove from the water, briefly rinse under cool water, and peel. Cut the eggs in quarters and use as a garnish, or use in specific recipes.

Homemade Sauerkraut

Makes about 3 quarts

For decades, Imwalle Gardens, a farm and fruit stand founded in Sonoma County, California, in 1886, made its own sauerkraut using huge cabbages—weighing up to twenty pounds each—shredded and then fermented in large oak barrels. Customers came from all over to purchase the sauerkraut, which was scooped out to order.

Although there had never been a problem with contamination, revised health department regulations required that the Imwalles switch to expensive stainless-steel containers; instead of changing, they quit making their sauerkraut and eventually the historic recipe was lost.

Today, commercial cabbage weighs a fraction of what it did then, when making sauerkraut was a common practice in most homes. You can still do it, though, and in small quantities if you like.

Recently, as home fermenting has become increasingly popular, ceramic fermenters fitted with ceramic weights that keep the cabbage submerged in brine are increasingly available at specialty shops, hardware stores and at some farmers markets. If you don't have one of these containers, you can put a plastic bag inside another and fit it with water. Tie it tightly, set a clean ceramic plate on top of the cabbage and set the filled bags on top; this will both weight the cabbage and seal the container.

A friend adds caraway seeds and sometimes grated carrots or grated beets to the cabbage when he makes sauerkraut; I generally add seasonings and other ingredients just before serving.

4 to 5 pounds white cabbage
1 ½ ounces (about ⅓ cup) kosher salt and more as needed

Remove any bruised outer leaves of the cabbage, core it, and slice it into very thin strips. In a large bowl, toss together the shredded cabbage and salt and let sit for 30 minutes.

Using very clean hands, massage the cabbage to release more juices. Alternately, pound the cabbage with a large wooden pestle.

Pack the cabbage and its juices into a glass or porcelain container; press down so that the juices completely cover the cabbage. Set the weight in place, cover the pot and set in a dark moderately warm part of the house. The cabbage will begin to ferment at between 68 and 70 degrees; it will take between 3 and 6 weeks for the fermentation to be complete. To test it, simply taste it. If it is pleasantly sour, it is done.

Use immediately, store in the refrigerator or pack into jars and process in a water bath (see Note about canning, page 273) and store in a dark pantry for up to 6 months.

Salted Mustard Greens

Makes about 1 pint

In the outdoor markets of Asia, there are often plastic tubs of greens preserved in brine—bok choy, cabbage, and sometimes greens hard for a Westerner to recognize. I came across a variation of this recipe, made with tangy mustard greens and the starchy liquid drained from cooked rice, in the delightful book Filipino Cuisine *by Gerry Gelle. You can serve these greens as a side dish, add them to soup, or chop them and stir them into rice, as on page 165.*

2 pounds mustard greens, washed and dried
½ cup kosher salt
¼ cup white rice

Combine the mustard greens and the salt in a large bowl, tossing them together well. Let the greens sit for an hour and then squeeze out all the liquid.

Rinse the white rice in running water until the water runs clear. Place the rice in a medium saucepan, add 4 cups of water, set over medium heat, and bring to a boil. Lower the heat, cover, and simmer for about 30 minutes. Strain the rice, saving the cooking liquid. Reserve the rice for another use and set the rice water aside to cool.

Pack the greens into a sterile pint glass jar. Pour the cooled rice water over the greens, cover with plastic wrap, and seal the jar. Age in a cool cupboard for 3 weeks.

Place in the refrigerator and use within six weeks.

Preserved Lemons

Makes 1 quart

Salty, tangy lemons, preserved in salt, vinegar, oil, or their own juice, have been a traditional element in North African cooking for centuries. They might be served alongside a savory tagine or minced and folded into the stew itself. They might be part of a platter of appetizers, to be enjoyed for their own distinctive flavor. When I wrote the first edition of this book, it was nearly impossible to find commercial preserved lemons. Now they are common. So why make them yourself? The best reason is if you have a lemon tree or access to fresh lemons, which I do.

I've experimented with every technique I've come across and this recipe, a combination of two methods, produces lemons I like a great deal. Try to find organic lemons and if you can't, wash and rinse the skins of your lemons thoroughly to remove any chemical residue. It is also important to realize from the start that the process takes a minimum of a week, so you can't begin in the morning and have preserved lemons for dinner, though if you're desperate, you can mince a clean, whole lemon, mix it with a tablespoon of kosher salt, set it aside, covered, for the day, and use it as you would preserved lemons in a recipe. Brining the fruit for several days before preserving them softens their skin, leaches out some of the pectin, and results in a slightly milder preserved lemon.

12 to 16 lemons, depending on size
1 cup kosher salt
1 tablespoon sugar
2 bay leaves, optional

Thoroughly wash between 8 and 12 lemons (the number will depend on their size, as many whole lemons as will, when cut into wedges, fill a quart jar), and place them in a large crock or glass bowl. (The remaining lemons will be used for juice.) Dissolve 1 tablespoon of the salt in a quart of water and pour it over the lemons. If it does not cover the lemons, mix more brine (using the same ratio of salt to water) and pour it over until they are covered. Cover the crock and set it aside for 1 day. (You can make the preserved lemons at this point, or you can cover them with new brine daily for up to 7 days.)

Rinse the brined lemons and dry them on a tea towel. Cut each lemon lengthwise into 6 to 8 wedges. Place them in a large bowl and top with ¾ cup of the remaining salt, the sugar, and the bay leaves, if using. Toss thoroughly and pack into a sterilized quart glass jar, tucking in the bay leaves about halfway through filling the jar, Juice the reserved lemons and pour the juice over the lemon wedges. If necessary, add fresh water until the lemons are covered completely. Cover the jar with heavy-duty plastic wrap and then seal with its lid.

Set the jar of lemons in a cool, dark cupboard for a week, giving it a good shaking once or twice a day. At the end of a week, move the jar to the refrigerator and use the preserved lemons within two or three months.

A Salt & Pepper Cookbook

Yogurt Cheese

Makes about 1 pound

Making cheese from yogurt is extremely easy, and produces very good results very quickly. It has the texture of cream cheese yet is tart, not sweet.

You will need clean cheesecloth to make this cheese, and you will need a place to suspend the bag while the whey drips out. I close the cheesecloth with a length of twine and then tie the twine to a kitchen cupboard doorknob, which works fine as long as I don't absentmindedly pull the door open. (Even then, the cheese is soft enough that it doesn't leave a bruise.) Use yogurt cheese as you would cream cheese—I find it especially good on toast—and to make Sikarni, page 313.

1 tablespoon fine sea salt, such as Velvet de Guerande or finely ground
   Himalayan pink salt
2 pounds plain whole milk yogurt

Stir the salt into the yogurt. Line a large strainer or colander with 4 layers of cheesecloth and pour the yogurt into it. Gather up with edges of the cheesecloth and tie it with twine to close to the yogurt. Suspend the bag over a bowl or other container to catch the whey; it should be away from direct sunlight or heat. Let the cheesecloth hang for between 24 and 36 hours, until no more whey drips from the cheese. Remove the cheese from its cheesecloth bag and store, covered, in the refrigerator. The cheese will keep for about 10 days.

Brined Pork Tenderloin

Makes 5 to 6 pounds

Meats cured in a brine of salt and sugar are called demi-sec; they are very easy to make successfully at home. Potassium nitrate, available at pharmacies, preserves the color of the meat; without it, the flesh turns gray but the taste is not altered; use it or not, it is up to you. Pork tenderloin is one of the easiest meats to prepare in this way because it is only about 1 ½ to 2 inches in diameter; it takes about a week to be fully cured. Larger cuts of pork, beef, or lamb require longer curing because it takes more time for the brine to penetrate them completely. Once the pork is cured, you can use it in a variety of recipes. It is excellent baked with a mustard, maple syrup, and black pepper glaze. Bean ragouts and potato gratins are good accompaniments. One tenderloin provides a generous serving for three people.

8 ounces kosher salt
8 ounces sugar
1 tablespoon potassium nitrate (saltpeter), optional
½ cup whole black peppercorns
2 tablespoons allspice berries
4 pork tenderloins, about 1 ¼ to 1 ½ pounds each

In a large pot, combine 8 cups spring water with the salt, sugar, and saltpeter, and set over high heat. Bring to a boil and stir until the sugar and salt are dissolved. Remove from the heat, add the peppercorns and allspice, and cool to room temperature.

Rinse the pork in cool running water and dry on a tea towel. Place the loins in a nonreactive container and pour the cooled brine over them. Use a clean plate or a small, clean cutting board to weight the tenderloins; they must be completely submerged in the brine. Cover tightly and place in the refrigerator or in a very cool pantry for 1 week. Discard the brine, wrap the pork, and use it within 10 days.

A Salt & Pepper Cookbook

Homemade Bacon

Makes about 6 pounds

1 fresh pork belly, about 6 pounds, skinless
1 ½ pounds kosher salt
4 ounces (¾ cup, packed) brown sugar
½ ounce cracked black peppercorns

5 or 6 fresh or dried bay leaves
2 tablespoons juniper berries, crushed
6 to 8 fresh thyme sprigs
Food-grade sawdust or wood chips, for smoking

Set the pork belly on a work surface. Find a glass, ceramic, or porcelain container into which it fits snugly. If you do not have such a container, cut the pork belly crosswise into pieces that will fit into the containers you have available.

In a large bowl, mix together the salt, sugar, and peppercorns and pour about half of it into the curing container. Rub the pork belly all over with some of the remaining mixture and press it into the container. Pack the remaining salt mixture on top so that the pork belly is completely covered by it. Cover with wax paper or parchment paper and set on a low shelf in the refrigerator.

Let the pork cure for 5 to 7 days. Remove it from the container, brush away all of the salt, and rinse the pork under cool water to remove any of the mixture that clings to it.

Set the pork belly on a tea towel set on clean, dry work, and pat it dry. If the slab is in one piece, cut it in half. Set one piece, fat side down, into a clean dry container. Set the bay leafs and thyme sprigs on top, and scatter the crushed juniper berries over the surface of the meat. Set the second piece of cured belly, fat side up, on top of the first piece. Cover with wax paper or parchment and let dry, in the refrigerator, for 1 to 2 days. Alternately, set the two pieces of belly on a rack (next to each other), and set an electric fan a few feet away, and turn it onto slow speed. Dry the bacon overnight.

Prepare a smoker according to the manufacturer's instructions. Smoke the bacon for a minimum of 8 hours and as long as 36 hours, depending on the degree of smokiness you prefer. If you are not sure when it is smoky enough, transfer a slab to your work surface, slice off a thin piece and taste it. If it isn't smoky enough, continue smoking it for several more hours.

Dry the smoked bacon in front of a slow-turning fan overnight, wrap it in parchment paper, and store it in the refrigerator. It will keep for 3 to 4 weeks, and is best during the first 2 weeks.

Vegetables

For decades, vegetables have been, for the most part, relegated to the side of the plate, to side dish status, something we eat now and then but often leave on our plates. That has changed as we have moved from a meat-and-potato culture to one that embraces the year's unfolding harvest in all its glorious diversity.

Salt and pepper are essential in helping us enjoy this bounty. Fresh corn, cooked just after being picked, needs nothing more than a bit of salt. Asparagus is never better than when fat stalks are roasted in a hot oven and seasoned with good salt and pepper.

When vegetables, such as artichokes and potatoes, are boiled, the water should be lightly salted to coax out flavor. When roasting root vegetables—solo or as a medley— salt added before cooking helps flavors blossom and pepper adds a delicious flourish that, when used with several vegetables, ties the dish together. Parsnips, carrots, beets, and celery root are especially delicious when cooked together in this way.

With more elaborate preparations, scalloped or mashed potatoes, a garlicky ratatouille, or a mushroom strudel, keep in mind that salt builds flavor, so add salt in stages for the best results. You'll end up using less than if you wait to add salt only at the end of cooking. If you want the full range of heat and flavor from pepper to infuse a dish, use white pepper during cooking and finish the dish with black pepper ground on the spot.

Roasted Asparagus

Haricots Verts with Zucchini Noodles, Sungold Tomatoes & Garlic

Glazed Carrots with Peppercorns

Black Pepper Zucchini

Basic Green Beans, with Variations

Grilled Corn with Butter & Pepper

Fall Succotash

Sweet Potatoes with Apple Cider & Black Pepper

Steamed Winter Squash with Black Pepper & Cilantro

Collard Greens with Ham Hocks & Maple Syrup

Brussels Sprout Leaves with Shallots & Bacon

A Salt & Pepper Cookbook

Roasted Asparagus

Serves 3 to 4

Once you've roasted asparagus, you may never boil or steam it again. It intensifies the flavor instead of diluting it, as the other techniques do, and you don't need to peel it. The only other method that rivals this one is grilling. In early spring, when the first asparagus appears, enjoy this as a delightful main course, followed by a green salad and, perhaps, with deviled eggs as an appetizer. As the season unfolds, use it as a base for more elaborate dishes.

16 to 20 fat asparagus stalks
Olive oil
Kosher salt
Black pepper in a mill

Preheat the oven to 450 degrees.

Snap off the tough stalks of the asparagus, set it on a baking sheet and drizzle with a little olive oil. Turn the asparagus in the olive oil so that each stalk is lightly and evenly coated. Season lightly with salt and set on the middle rack of the oven and cook until tender, about 5 to 6 minutes for thin stalks and up to 15 for the fattest stalks.

Remove from the oven, transfer to a serving platter and season with several generous turns of black pepper. Enjoy immediately.

Variations:

- Top with sieved hard-cooked egg before adding the black pepper.
- Divide among individual plates, top with a poached egg and your favorite warm vinaigrette.
- Cut the asparagus into 2-inch diagonal pieces before roasting it. After cooking it, fold it into a simple risotto and or toss it with a medium pasta, such as gemelli or strozzapreti, along with fresh favas, grated lemon zest, good olive oil, and, of course, plenty of freshly ground black pepper. Add a shower of grated pepato just before serving.

Haricots Verts with Zucchini Noodles, Sungold Tomatoes & Garlic

Serves 4

I originally made this quick summer dish using pasta but as so many friends have, for a variety of reasons, stopped eating wheat, I developed this dish, which uses zucchini in place of spaghettini. It is simple, delicious, and can be either a main course or a side dish. For a more substantial main course, cook 2 sausages (andouille, Spanish chorizo, or linguica), slice them into half moons and add them to the zucchini with the tomatoes.

Kosher salt
6 ounces haricots vert, trimmed
2 medium zucchini
4 tablespoons extra virgin olive oil
4 garlic cloves, minced
Black pepper in a mill
1 pint golden cherry tomatoes,
preferably Sungold variety, halved or quartered
2 tablespoons fresh herbs: a mix of Italian parsley, chives, basil, summer savory, and oregano
3 ounces ricotta salata, grated

Fill a medium saucepan half full with water, season generously with kosher salt and bring to rolling boil. Add the haricots verts and when the water returns to a boil cook for 1 minute; test for doneness and cook for 30 seconds more if needed. Drain thoroughly.

Meanwhile, use a mandoline to cut the zucchini into spaghetti-sized ribbons.

Pour the olive oil into a large saucepan, set over medium heat, add the garlic and cook for about 30 seconds. Add the zucchini, toss and cook until it just loses its raw texture, about 2 to 3 minutes.

Season generously with salt and pepper.

Add the tomatoes, toss and cook for 1 minute more. Add the drained haricots verts and the herbs, toss and transfer to a warmed serving bowl.

Add the cheese, toss again, taste for seasonings and serve right away.

Glazed Carrots with Peppercorns

Serve 4 to 6

Pepper highlights virtually all sweet flavors and goes beautifully with carrots. Here, additional elements are contributed by green peppercorns, with their fresh, mildly sour notes, and the aromatic toastiness of Sichuan peppercorns.

Kosher salt
¾ pound carrots, preferably medium-sized Nantes variety, trimmed and
    peeled or scrubbed
4 tablespoons butter
3 tablespoons brown sugar
2 tablespoons apple cider vinegar
1 teaspoon coarsely ground black pepper
½ teaspoon crushed green peppercorns
½ teaspoon Sichuan peppercorns, toasted, crushed, and sifted (see Note,
    page 271)
1 teaspoon kosher salt

Fill a medium saucepan half full with water, season with about a tablespoon of kosher salt and bring to a boil over high heat.

Cut the carrots into thin diagonal slices.

Blanch the carrots for 90 seconds, drain thoroughly and set aside briefly.

Empty the saucepan, dry it with a tea towel and set it over medium heat. Add the butter and sugar; when the butter is melted and the sugar dissolved, add the carrots to the pan and toss to coat them thoroughly.

Add the vinegar, reduce the heat to low, cover the pan, and simmer very gently until the carrots are tender, about 7 or 8 minutes or a tad longer.

Add the peppercorns, toss gently, season with salt, tip into a serving bowl and serve right away.

Black Pepper Zucchini

Serves 4 to 6

As soon as the first zucchini appear in late spring or early summer, I make this dish, which is so much more than the sum of its parts. It is ridiculously delicious and anyone, even the most timid cook, can make it successfully.

If you have a garden, feel free to use any mixture of summer squashes, though I find it best with very young zucchini that has a greater ratio of skin to flesh than larger ones. Do not be tempted to skimp on the quantity of black pepper, and use the best you have; it's the crucial ingredient.

3 tablespoons clarified butter
6 fresh zucchini, about 6 inches long, trimmed and sliced ⅛ inch thick
2 tablespoons coarsely crushed black pepper
Kosher salt

Heat the clarified butter in a large heavy skillet set over medium heat. Add the zucchini and sauté, tossing frequently, until the zucchini is just barely tender, about 4 minutes. Add the black pepper, season with salt, remove from the heat, and serve immediately.

Basic Green Beans, with Variations

Serves 4

Whenever you cook fresh green beans, you should invite Julia Child to sit on your shoulder and whisper in your ear. She was adamant that green beans need proper cooking for their flavors to emerge and was disgusted by raw or undercooked green beans. One afternoon I was at a special luncheon honoring Julia and her colleague Jacques Pepin. When the main course arrived, Jacques, who was seated next to me, leaned over and whispered, "Thank God the green beans are cooked!" We both giggled and glanced at Julia, who was seated nearby.

This recipe works with any type of green bean except tiny haricots verts, which should be cooked for just 60 to 90 seconds, depending on their size. Be sure to test them before removing them from the heat. If you ever come across Spanish Musica, a variety of long, flat green beans, snag them. Cooked for just 3 minutes and slathered in butter, they are delicious enough to make a full meal.

Kosher salt
1 ¼ pound green beans, trimmed (see Note on next page)
3 tablespoons butter, preferably organic
Flake salt
Black pepper in a mill

Fill a large pot two-thirds full with water, add a tablespoon of salt for every 2 quarts and bring to a boil over high heat. When the water reaches a full rolling boil, add the green beans, stir and watch until the water returns to a boil. Cook for 3 to 4 minutes, depending on the vareity of bean, remove a bean and carefully taste it for doneness. Continue to test every 30 seconds until the beans are perfectly cooked; your teeth should bite through easily but with the tiniest bit of resistance.

Immediately drain the beans thoroughly, tip into a warmed serving bowl, add the butter toss and season with flake salt and several turns of black pepper. Serve immediately.

NOTE

The stem end of green beans should be snipped off but the blossom ends do not need to be removed. The easiest way to do this is to arrange the beans on your work surface with the stem ends lined up evenly together. Use a sharp knife to cut close to where the stem connects with the bean.

Variations:

- Cut 10 to 12 large basil leaves into very thin ribbons and add to the beans with the butter.

- Add 1 tablespoon chopped Italian parsley to the beans with the butter and squeeze the juice of ½ lemon over the beans before seasoning with salt and pepper.

- Fry 3 or 4 strips of bacon until crisp; drain the bacon and, when it is cool, crumble it. Pour off all but 2 tablespoons of the bacon fat and sauté a minced shallot in the fat that remains. Remove from the heat and add 2 teaspoons of lemon juice or best-quality white wine vinegar. When the green beans are cooked, transfer them to a warm serving bowl, add the shallot mixture, season with salt and pepper and top with the crumbled bacon.

Grilled Corn with Butter & Pepper

Serves 4 to 6

In June, 1998, during my first visit to Malaysia, I walked from my hotel in Kuala Lumpur to the city's Night Market, something akin to our Wednesday Night Market in downtown Santa Rosa, but much bigger.

I walked along the rows, attracting curious looks and searching for the most interesting foods, preferably things I didn't recognize or had never eaten. The aromas hovering around one stall were particularly evocative and I was surprised to discover that they offered a single item, corn on the cob, each hot ear plucked from an enormous cauldron of simmering water and covered with the finely ground white pepper that is common on every Malaysian table.

I ponied up the equivalent of about fifty cents, took my corn and walked to a nearby stoop to eat. I was nibbling the most delicious corn I've ever tasted, so fresh and rich that it needed no butter to enhance it. Before the night was over, I ate three, maybe four, ears. I ate nothing else that evening, headed back to my hotel and soon drifted into a luxurious, satisfied sleep.

Here is my version of corn on the cob inspired by this visit. If you grow your own corn, you'll get closer to what I enjoyed in that far away city by bringing a pot of salted water to a boil and then picking and shucking the corn so that you can plunge it into the boiling water seconds after it is off the stalk. Pull it out of the water bath after just 2 to 3 minutes, shake off excess water and sprinkle it all over with finely ground white pepper.

½ cup (1 stick) unsalted butter
½ to 1 teaspoon ground white pepper
1 teaspoon crushed black peppercorns
1 teaspoon kosher salt
6 fresh ears of corn, shucked

Melt the butter in a small saucepan over medium heat. When it is melted, remove it from the heat and stir in the peppers and salt. Set aside and keep warm.

Grill the corn over a charcoal fire or on a stove-top grill. When it is done, transfer to a serving platter. Using a pastry brush, brush each ear of corn generously with the pepper butter. Serve immediately, with the remaining butter alongside.

Fall Succotash

Serves 3 to 4

    Succotash is a traditional Native American dish, typically made with what is known as the "three sisters," corn, shell beans, and squash. This version is one of many that I make and calls for the poblanos that are abundant in the fall where I live. Later in the fall, I use shell beans and winter squash. Although succotash is typically served as a side dish or as a bed for roasted salmon, roasted chicken, or grilled beef, it is also a wonderful main course on a warm fall night.

3 tablespoons butter
1 medium shallot, minced
2 to 3 ounces pancetta or guanciale, in small cubes
2 garlic cloves, minced
2 poblanos, seared, peeled, seeded, and cut into small dice
Kernels from 4 fresh ears of corn
Flake salt
Black pepper in a mill

Put the butter into a medium sauté pan set over medium-low heat and, when it is melted, add the shallot and sauté gently until soft and fragrant, about 7 to 8 minutes. Add the pancetta or guanciale and continue, stirring now and then, until it loses its raw look.

Add the garlic, poblanos and corn, season lightly with salt, cover, reduce the heat and cook very gently for 4 to 5 minutes, until the corn just loses its raw taste. Remove from the heat, season generously with black pepper, taste and correct for salt.

Sweet Potatoes with Apple Cider & Black Pepper

Serves 4 to 6

Sweet potatoes, like carrots, are perfectly accented by black pepper. These tubers are not related to potatoes, but rather are members of the morning glory family and native to the American tropics. They are not, as is often assumed, yams, though they are frequently sold under that name. Yams belong to a different botanical species entirely and are native to Asia and Africa; they are huge and starchy, with a single tuber often weighing as much as a hundred pounds.

2 ½ to 3 pounds (about 2 large) sweet potatoes, peeled and sliced ⅛ inch thick
1 cup apple cider or juice
3 tablespoons apple cider vinegar
2 teaspoons freshly ground black pepper
Flake salt

Place the sweet potatoes in a wide saucepan, add water to barely cover, bring to a boil over medium heat, reduce to low, and simmer, covered, until tender, about 10 to 15 minutes. Drain and discard the water, return the potatoes to the heat, and add the apple cider, vinegar, and black pepper. Sauté, stirring frequently and carefully, over low heat until the apple cider is reduced to a glaze. Transfer to a serving bowl, season with salt, and serve immediately.

Steamed Winter Squash with Black Pepper & Cilantro

Serves 3 to 4

You can use any hard-skinned winter squash for this dish, but if you can find Tahitian Melon squash, which is dense-fleshed and intensely flavored, use it. These large squash can weigh ten pounds or more, and have an elongated neck extending from a plump round body. The neck is solid meat and is excellent sliced and steamed; reserve the body for pies and cookies and use the neck in this recipe. The best place to find Tahitian Melon and other uncommon winter squashes is at a farmers market; farmers often sell portions of larger squash, but if you must purchase an entire one, you can steam it and freeze it for later use.

1 ½ pounds hard winter squash, such as Tahitian Melon squash
3 tablespoons butter
1 tablespoon freshly cracked black pepper, plus more to taste
Flake salt
2 tablespoons fresh minced cilantro leaves

Cut the squash in half lengthwise and use a very sharp paring knife to peel it. Cut the squash into slices not quite ¼ inch thick, place them in the top of a steamer, set over boiling water, cover, and steam until just tender, about 7 or 8 minutes.

Melt the butter in a large frying pan, add the black pepper and add the steamed squash, and toss gently to coat the squash evenly with butter and pepper.

Season with salt, tip into a serving dish, taste, correct the seasoning, add the cilantro and serve.

Collard Greens with Ham Hocks & Maple Syrup

Serves 4 to 6

Steve Garner, the host of "The Good Food Hour" on KSRO-AM in northern California, is a modest yet sensational cook. Like all the best cooks, Steve loves to eat and it is his robust passion that underscores his signature dishes, such as these classic southern collard greens-with-a-twist. His credentials for greens are intact (he's from Louisville, Kentucky), so when he says that the smoky flavor of the ham hocks adds an element lacking with traditional salt pork, you can trust him. The inspired ingredient, though, is the maple syrup, which is absolutely irresistible with the tangy, salty, peppery flavors in this homey dish.

2 pounds collard greens
2 tablespoons olive oil
2 tablespoons peanut oil, plus more as necessary
3 or 4 ounces smoked ham–hock meat, minced
Kosher salt
¾ cup unsalted chicken broth or stock
3 tablespoons apple cider vinegar
Black pepper in a mill
½ teaspoon crushed red pepper or several shakes of Tabasco sauce
1 tablespoon maple syrup

Wash the collard greens and shake them to remove some but not all of the water that clings to their leaves. Cut away the stems, stack the leaves, and roll them up (like a big cigar). Slice them into ¼-inch strips.

In a large saucepan, heat the olive oil and peanut oil, add the minced ham-hock meat, and sauté over medium heat for 2 or 3 minutes. Add a handful of greens, use a wooden spoon to push them down into the pot, and sauté until they wilt; repeat, adding a handful of greens at a time and cooking until they wilt before making the next addition. Stir the greens, and if there isn't a slight glisten from the oil, add another tablespoon. Season with a generous pinch of coarse salt, add the stock, and when it begins to boil, reduce the heat as low as possible. Cover and simmer until the greens are tender but not mushy, about 15 minutes. Check now and then to be certain the liquid has not evaporated; if the pan gets dry before the greens are tender, add a little more broth or stock. When the greens are done, remove the lid, increase the heat, and evaporate any remaining liquid. Add the vinegar, several turns of black pepper, and the pepper flakes or Tabasco. Stir the mixture and pour the maple syrup over the greens. Stir again, quickly, remove from the heat, place the greens in a bowl, and serve immediately.

Brussels Sprout Leaves with Shallots & Bacon

Serves 8 to 10

This dish wins converts every time I serve it and it has become a standard addition to my Thanksgiving table. People who say they hate Brussels sprouts reach for seconds and sometimes thirds and the dish is often more popular than the main course it accompanies. Always make more than you think you'll need for a single meal so that you'll have leftovers.

Try to find Brussels sprouts that are still on their stalk, as they will be fresher and will not have the unpleasant flavors that old Brussels sprouts develop. The time to find them is in mid fall through early winter.

1 stalk of Brussels sprouts
8 to 10 bacon slices, diced
2 shallots, minced
Flake salt
Black pepper in a mill

Cut or snap each sprout off the stalk; pressing outward with your thumb usually works.

Use a sharp pairing knife to cut each sprout in half through its poles and then cut out the tiny core. Discard the cores or, if you or a neighbor keep chickens, save them for the birds.

Separate the leaves of each sprout. This will take a while and you'll have a better time if you listen to music, watch a video, or sit in the garden while you do it. This can be done a day before cooking; cover and refrigerate the leaves until ready to use.

To finish the dish, sauté the bacon in a pot big enough to hold all the leaves until it is almost crisp. Add the shallots and sauté until soft and fragrant, about 5 minutes; do not let the shallots brown. Add the Brussels sprout leaves, cover the pan and cook for 3 to 4 minutes, until the leaves just begin to wilt a bit. Uncover and continue to cook, turning the leaves frequently so that they bacon and shallots are evenly distributed. As they cook, they will continue to wilt and the volume will decrease substantially.

Continue to cook and stir until the leaves are tender, about 10 minutes. Season to taste with salt and pepper, transfer to a bowl and serve.

What to do with leftovers: These Brussels sprouts are delicious with pasta. Simply heat them through as you cook a small pasta—orecchiette is ideal—and then toss the two together. Add a splash of extra virgin olive oil and a shower of grated cheese such as dry Jack or pepato. They are also wonderful folded into a simple risotto moments before it is done cooking.

Sweets

When I was a child I often saw friends reach for a salt shaker whenever we shared watermelon or cantaloupe. The salt made it sweeter, they always said, and although I didn't agree—salt made the fruit salty—I did like a few grains on a Pippin apple.

As I worked on the first edition of this book, a friend from the Midwest told me that her father taught her to sprinkle pepper over slices of cantaloupe or other muskmelon. She still swears by the combination, which works equally well on pineapple, pears, papaya, mango, and, especially, strawberries.

Using salt and pepper to complement the flavors of fruit and other sweets has a long history. Salt is used to boost the flavors of countless desserts; a few sweets, saltwater taffy, for example, rely on saltiness in addition to sweetness for their characteristic tastes.

Adding white or black pepper to desserts was commonplace in Roman and Medieval times. Although the practice had all but disappeared in the early twentieth century, it began to reemerge as jalapeños, serranos, and other hot chiles made their way into desserts. Increasingly, professional chefs and bakers began to discover the suave, subtle characteristics of the peppercorn, which enhances rather than dominates a sweet dish, and desserts with pepper are no longer unusual in restaurants. By the time I began working on the revised edition, salt, especially, had found its way into chocolate, everywhere.

In the last decade or so, bacon has started to appear in sweets from toffee, divinity and chocolate truffles to all manner of cookies, doughnuts, s'mores, cupcakes, cakes, pies, and more.

Roasted Strawberries with Black Pepper

Grilled Peaches with Black Pepper & Salted Caramel Sauce

Pineapple Granita with Black Pepper

Sikarni

Black Pepper Ice Cream

Salt & Pepper Shortbread

Adult Sugar Cookies

Salt Brownies

Salted Double Ginger Brownies

Peppered Carrot Cake with Salted Cream Cheese Frosting

Roasted Strawberries with Black Pepper

Serves 4 to 6

I love the pristine taste of these strawberries straight out of the oven but you can create a more elaborate dessert by adding a spoonful of creme fraiche or mascarpone, Black Pepper Ice Cream, page 314, Salt & Pepper Shortbread, page 315, or Adult Sugar Cookies, page 316.

2 pint baskets strawberries, stems removed
3 tablespoons sugar
Pinch of kosher salt
1 tablespoon freshly ground black pepper

Rinse the strawberries in cool water, place in a strainer or colander, and shake off most of the water. Slice the strawberries about ⅛ inch thick, place them in a large bowl, and sprinkle them with the sugar. Cover and refrigerate for at least 1 hour and up to 4 hours.

Preheat the oven to 375 degrees.

Add the salt and the black pepper, stir once gently and transfer to an oven-proof baking dish. Roast for 8 to 10 minutes, until the juices are bubbling and the strawberries are hot but not mushy. Divide among individual dishes and serve immediately.

Grilled Peaches with Black Pepper & Salted Caramel Sauce

Serves 3 to 6

I rarely suggest purchasing a commercial product when you can easily make something at home but the caramel sauce here is an exception. If you happen to have wonderful ripe peaches and no time to make the sauce, there are excellent caramel sauces on the market these days, some salted, some not. It is a simple thing to salt a commercial caramel sauce to taste.

This lovely dessert is best made when you have an outdoor charcoal or wood fire, though you can, of course, make it in a stovetop grill or grill pan.

Salted Caramel Sauce (recipe follows)
6 ripe peaches, preferably from an orchard near you
Half a lemon
Black pepper in a mill

Warm the caramel sauce and set it aside.

Cut the peaches in half through their poles, remove the pips and squeeze a little lemon over the cut portions of the peaches.

If cooking over charcoal or wood, wait until the fire has died down to glowing embers. Otherwise, heat a stovetop grill or grill pan over medium heat.

Add the peaches in a single layer, cut side down, and cook for about 5 minutes, until the fruit just begins to soften and takes on grill marks. Use tongs to turn over and cook for 4 or 5 minutes more. Transfer to a plate and season all over with black pepper.

Set on individual plates, drizzle with warm caramel sauce.

Variation:

Serve with a scoop of black pepper ice cream or other ice cream.

Salted Caramel Sauce

Makes about 1 ½ cups

1 cup granulated sugar
1 stick (4 ounces) butter, preferably organic, in small pieces, at room
  temperature
½ cup heavy cream
2 teaspoons fleur de sel, sel gris, or smoked salt, plus more for sprinkling

Put the sugar into a small saucepan with a heavy bottom set over
medium-high heat. As the sugar dissolves, do not stir it or agitate the pan.
When the melted sugar takes on a golden color, remove the pan from the
heat and carefully whisk in the butter, a piece at a time, being careful not
to burn yourself.

Slowly pour in the cream, whisking all the while, and add the 2 teaspoons of
salt with the final addition of cream. Continue to whisk until the sauce cools
to room temperature.

Pour the cooled caramel sauce into a glass jar or, if using right away, pitcher
and sprinkle lightly with salt. The sauce will keep—ha!—for 2 to 3 weeks in
the refrigerator.

To warm, set the jar in a small saucepan half full with water and set over
low heat.

Use in any recipe that calls for caramel sauce.

Pineapple Granita with Black Pepper

Serves 4 to 6

The inspiration for this dish comes from my visit to Sarawak, one of the thirteen states of Malaysia. Looking through the cookbook Sarawak Pepper, *published by the Pepper Marketing Board of Malaysia, I saw a photograph of pineapple studded with black pepper and knew instantly how wonderfully the flavors would fit together. That night, I scattered some crushed black pepper over sliced pineapple and my suspicions were confirmed. My recipe is adapted from one in the book.*

2 cups fresh ripe pineapple, minced
Black pepper in a mill
1 cup Simple Syrup with Pepper (page 365)
Juice of 2 lemons
Sprigs of fresh mint

Put the pineapple into a bowl, season with several turns of black pepper, stir in the simple syrup and lemon juice, place in a shallow stainless steel pan, and put it in the freezer until it is nearly frozen. Remove from the freezer, mix with a fork to break it up, and return to the freezer until it is completely frozen.

To serve, put generous scoops of granita into chilled glasses or bowls. Garnish with a mint leaf and serve immediately.

Variation:

Before putting the granita into glasses, add several slices of pineapple, seasoned with salt and pepper.

Nepalese Sikarni

Serves 6 to 8

Sikarni hails from Nepal and is simple to make anywhere there is good yogurt. Sometimes, it is the texture of custard; other times it is as thick as cream cheese. Ingredients vary, too, though yogurt, spices, and sugar form the foundation. Some versions incorporate fruit but I prefer to serve fresh fruit alongside, rather than mixed in.

⅛ teaspoon saffron threads
1 tablespoon whole milk
2 teaspoons cracked black peppercorns
½ teaspoon ground cinnamon
¼ teaspoon ground cardamom
1 ¼ cup yogurt cheese (page 283), at room temperature
1 cup granulated sugar
1 tablespoon grated fresh coconut
2 tablespoons raisins
2 tablespoons shelled pistachios
1 tablespoon chopped dates

Put the saffron threads into a small mortar, add the milk and crush the saffron. Add the peppercorns, cinnamon, and cardamom and set aside.

Put the cheese and sugar into a medium bowl and mix by hand until smooth. Fold in the coconut, raisins, pistachios, dates, and the milk mixture and mix until smooth.

Cover and refrigerate until thoroughly chilled. Serve with fresh fruit alongside.

Serving Suggestions

- with sliced or cubed mango
- with sliced papaya
- with fresh pomegranate arils
- with sliced pineapple seasoned with salt and pepper

Black Pepper Ice Cream

Makes about 6 cups

Ice cream spiked with black pepper began to appear in American restaurants in the1990s but it has ancient roots. Two thousand years ago, pepper, along with many other spices, was a common ingredient in custards and other desserts.

4 cups half–and–half
1 tablespoon black peppercorns, coarsely cracked
Half a vanilla bean, split open
¾ cup sugar
4 egg yolks
2 teaspoons freshly ground black pepper
½ teaspoon kosher salt

Scald the half-and-half in a medium saucepan over medium-high heat. Remove from the heat, add the cracked peppercorns and vanilla bean, cover, and let the mixture steep for 30 minutes. Strain into a clean saucepan, discard the peppercorns, and set aside the vanilla bean (it may be rinsed, dried, and used again).

In a medium bowl, whisk together the sugar and egg yolks, whipping vigorously until the mixture is pale yellow and forms a ribbon when dropped from a spoon. Stir in the ground pepper and salt, and whisk the mixture into the half-and-half. Set the saucepan over medium-low heat and stir constantly until the custard thickens.

Remove from the heat immediately and whisk until the mixture cools slightly. Cover tightly with plastic wrap and refrigerate until thoroughly chilled.

Freeze in an ice cream maker according to the manufacturer's instructions. Store, tightly covered, in the freezer until ready to serve.

Salt & Pepper Shortbread

Makes 16 wedges

This shortbread is based on the traditional shortbread of Scotland; it is rich and slightly moist and delicately perfumed by the warm peppercorns. I originally developed this recipe with lavender for Matanzas Creek Winery, located in Bennett Valley in southeast Santa Rosa, California. The winery has an extraordinary two-acre lavender field at its entrance and has celebrated the aromatic flower in a variety of ways over the years.

½ cup superfine sugar
1 teaspoon freshly ground black pepper
1 teaspoon freshly ground white pepper
1 teaspoon kosher salt
8 ounces butter, preferably organic, cut into cubes and chilled
½ teaspoon vanilla
2 cups flour

Put the sugar, peppers, and salt in the work bowl of a food processor fitted with the metal blade and pulse several times. Add the butter and pulse until the sugar and butter are smoothly blended; add the vanilla and pulse again. Add the flour and pulse, stopping two or three times to scrape the work bowl with a rubber spatula, until the mixture is evenly mixed and crumbly. Cover your work surface with plastic wrap and transfer the mixture to it; use your hands to press the dough together, then knead it gently until it just holds together. Wrap it tightly and refrigerate for 30 to 60 minutes.

Preheat the oven to 250 degrees.

Divide the dough in half and press each half into an 8-inch tart pan, an 8-inch cake pan, or an 8-inch glass pie pan. Set on the top rack of the oven and bake for about 20 minutes. Rotate the pans so the shortbread cooks evenly and continue to cook for 30 to 40 minutes more, until the shortbread is just barely showing a little color.

Adult Sugar Cookies

Makes 6 to 7 dozen

These cookies are too spicy for most children's palates, but adults love the bright, sharp kick provided by a blend of salt, pepper, and spices.

1 ½ cups butter, at room temperature
1 cup sugar
2 whole large eggs or 4 egg yolks, beaten
1 teaspoon vanilla extract
2 teaspoons grated fresh ginger
3 ¾ cups all-purpose flour
1 tablespoon kosher salt
1 teaspoon finely ground white pepper
1 teaspoon finely ground black pepper
2 teaspoons hot mustard flour (such as Colman's)
1 teaspoon ground ginger
1 teaspoon ground cayenne or other ground hot chili
1 egg white
Red Hot Sugar (see Note on next page)

Using a heavy-duty whisk or an electric mixer, cream together the butter and sugar. Add the eggs, vanilla, and fresh ginger, and mix thoroughly. In a separate bowl, combine the flour, salt, pepper, mustard, ground ginger, and cayenne. Add the flour mixture, one half at a time, to the butter mixture, and mix together thoroughly. Press the dough into a ball, wrap it in plastic, and chill for at least 2 hours.

To make the cookies, remove the dough from the refrigerator 30 minutes before rolling it out.

Preheat the oven to 350 degrees.

Cut the dough into 3 equal pieces. Use the palms of your hands to roll out the dough on a lightly floured surface until it forms a rope about 1 ¼ inches in diameter. Using a very sharp knife, slice the dough into ⅜-inch-thick rounds and set them about 1 inch apart on an ungreased baking sheet.

Mix the egg white with 2 tablespoons of water and brush the surface of each cookie lightly with the mixture and sprinkle the Red Hot Sugar on top. Bake for 7 to 9 minutes, until the cookies just barely begin to color. Remove from the oven and cool on a rack. Serve immediately or store in an airtight container for up to 2 weeks.

NOTE

To make 1 cup Red Hot Sugar, combine 1 cup granulated sugar, 1 teaspoon kosher salt and ¼ teaspoon (and more to taste) cayenne and place in a container with a lid. Add several drops of red food coloring, close the container, and shake it until the sugar is evenly colored. Repeat until the sugar is shade of red you want. Colored sugar will keep indefinitely in a tightly sealed container.

Salt Brownies

Makes about 2 dozen

Either I did not inherit the chocolate gene or something traumatic happened early in my life to make me at best tepid about it. I'd never much cared for it and then my friend Rosemary made a platter of brownies for my birthday a few years ago. I felt obligated to take a bite before I gave them away and what a revelation that bite was. Understanding my passion for salt, Rosemary sprinkled the top of the brownies with a good measure of it before baking them. I confess that I ate them all.

I am hardly alone. Salted chocolate is a trend that is sweeping the confectionery industry. Small boutique chocolatiers, such as Gandolf's Fine Chocolates and Recherche du Plaisir, both located in Sonoma County where I live, count salted chocolates and salted caramels among their best sellers. Large companies such as Trader Joe's also offer salted chocolate and caramels, sometimes together in a single, irresistible bar. You have been warned!

1 tablespoon plus 6 ounces (1 ½ sticks) butter, preferably local and organic
2 ounces best-quality unsweetened chocolate, grated or chopped
2 cups granulated sugar
2 teaspoons kosher salt
3 large eggs from pastured hens, at room temperature
1 teaspoon pure vanilla extract
1 cup unbleached all-purpose flour
6 tablespoons best-quality unsweetened cocoa
1 tablespoon, or to taste, Maldon salt flakes

Preheat the oven to 350 degrees.

Use the tablespoon of butter to coat the inside of a 9-inch square baking dish. Set aside.

Fill the bottom part of a double boiler with a couple of inches of water and set over high heat. When the water begins to simmer, reduce the heat to very low. Put the 6 ounces of butter and the grated chocolate into the top part of the double boiler and set it over the simmering water until both are melted. Remove from the heat and let cool.

Put the eggs into a mixing bowl and beat them (with a whisk or electric mixer) until they are pale yellow and quite foamy. Slowly add the sugar and kosher salt; beat until creamy.

Use a rubber spatula to quickly fold in the cooled chocolate and butter mixture and the vanilla; do not over mix at this point.

Combine the flour and cocoa and fold into the chocolate mixture; again, do not over mix.

Pour the batter into the buttered baking dish and agitate the dish slightly so that the batter settles evening. Sprinkle the salt flakes over the batter, set on the middle rack of the oven and bake for 25 minutes.

To test for doneness, insert a toothpick into the center and remove it; if done, it will come out almost but not quite clean. If there is a lot of raw batter on the toothpick, cook for another 5 to 10 minutes.

Remove the brownies from the oven and let cool.

Cut into squares and enjoy.

Salted Double Ginger Brownies

Makes 18

The "Ginger" in the title of this recipe refers to both the flavor of fresh ginger and to their color. "Brownies" without chocolate are often called "blondies" but, really, ginger makes more sense, especially now that South Park *has popularized the use of the term to describe anyone with red hair.*

A version of these butterscotch brownies—another common name—from Betty Crocker's Cookbook for Boys and Girls *is the first recipe I ever followed, back when I was 8 years old and a neighbor gave me the book for my birthday. All these years later I still love them, more so since I began adding fresh ginger to the batter and sprinkling salt on top.*

1 tablespoon plus ¼ cup unsalted butter, preferably organic
¾ cup all-purpose flour
1 teaspoon baking powder
2 teaspoons kosher salt
White pepper in a mill
1 cup, packed, light brown sugar
1 large pastured egg
2 teaspoons grated fresh ginger
1 teaspoon vanilla extract
¾ cup lightly toasted and chopped walnuts
2 teaspoons Maldon Salt Flakes

Preheat the oven to 350 degrees.

Use the tablespoon of butter to coat the inside of a 9-inch square pan. Set it aside.

Combine the flour, baking powder, and kosher salt in a small bowl, add several turns of white pepper, mix with a fork and set aside.

Put the butter in a small saucepan set over very low heat and when it is melted, stir in the sugar.

Put the egg into a medium mixing bowl and whisk vigorously until it is light and creamy. Fold in the butter mixture, the ginger, and the vanilla.

Use a rubber spatula to fold in the flour mixture and the nuts, if using. Pour into the buttered baking pan, using the spatula to remove all of the batter from the mixing bowl.

Sprinkle the salt flakes over the batter and bake for 20 to 25 minutes, until lightly browned on top but still a tad soft in the center.

Remove from the heat, cool slightly, cut into squares and finish cooling on a rack.

Peppered Carrot Cake with Salted Cream Cheese Frosting

Serves 6 to 8

One of my first restaurant jobs was in a little diner called Jerome's Good Dogs. After working there for a few months, I was promoted to manager, which involved not just supervising the staff and placing all orders but also making our salads, sauces, and desserts. I'd make six large carrot cakes and six large chocolate cakes every other day. The café is now part of Cotati, California, folklore and Jerome, too, is sadly gone. But the memory of his carrot cake lives on and I suspect he would be appalled by how I have transformed it to satisfy my perverse palate and its craving for salt and spice.

2 teaspoons butter
2 cups unbleached all-purpose flour, sifted
1 teaspoon baking powder
1 teaspoon baking soda
2 teaspoons freshly ground black pepper
1 teaspoon freshly ground white pepper
1 teaspoon kosher salt
1 teaspoon ground cinnamon
1 teaspoon ground cardamom
4 large farm eggs
1⅓ cup mildly flavored olive oil or cold-press grapeseed oil
1 cup granulated sugar
¾ cup, packed, light brown sugar
1 pound carrots, peeled, trimmed, and grated on a large blade
1 cup chopped walnuts or pecans, lightly toasted, optional
Salted Cream Cheese Frosting (recipe follows)
½ teaspoon flake salt, such as Maldon Salt Flakes or Murray River Salt

Preheat the oven to 375 degrees.

Coat the inside of a 10-inch spring-form pan with the butter. Set it aside.

Put the flour, baking powder, baking soda, peppers, salt, cinnamon, and cardamom into a large mixing bowl and stir well with a fork. Set aside.

Put the eggs into a large mixing bowl and whisk well. Add the oil and both sugars and whisk until smooth.

Using a rubber spatula, tip the sugar mixture into the dry ingredients and mix together quickly. Fold in the carrots and the nuts, if using.

Pour into the buttered pan.

Set the pan on the middle rack of the oven and bake for 35 to 40 minutes, until the cake is golden brown and the center springs back when very lightly touched. Remove from the oven, set on a rack and cool for 15 minutes before releasing the spring form.

While the cake cooks, make the frosting.

When the cake has cooled to room temperature, spread the frosting on top. Sprinkle the flake salt over the frosting and serve right away.

Leftover cake will keep in the refrigerator for 2 to 3 days.

A Salt & Pepper Cookbook

Salted Cream Cheese Frosting

8 ounces old-fashioned cream cheese, at room temperature
1 stick organic butter, at room temperature
1 cup sifted confectioner's sugar
¾ teaspoon kosher salt
¼ teaspoon finely ground black pepper
¾ teaspoon vanilla extract

Put the cream cheese in a medium bowl and mix with a sturdy wooden spoon until smooth. Add the butter, a big dollop at a time, mixing between additions. When the mixture is smooth, mix in the sugar.

Put the kosher salt and the black pepper in a little pile near the edge of the bowl, pour the vanilla extract on top and gently agitate the bowl so the salt dissolves. Mix well.

Cover and refrigerate until the cake comes out of the oven.

Beverages

During 1998 several of my friends visited Tibet. Each time one returned, I asked: "How was the tea?" Usually, they would mumble something about being tired of yak butter and change the topic to the spectacular landscape or vanishing culture. It was hard to get a direct answer.

"You certainly drink gallons of it," Forrest Tancer, former winemaker at Iron Horse Vineyards, commented shortly after his return from a long trek, "and you're really happy to have it after hiking."

The tea in question is a mild black tea favored with yak butter and salt. I've had it in the United States, and must confess that I found little to recommend it, though I suspect this is because of context. Traditional beverages exist not only because certain ingredients are available in a specific location, but also because those ingredients fill certain needs, in Tibet, for instance, the need for energy and warmth, the need for sufficient fat, the need for salt. Tibetan-style tea, with butter made from cow's milk rather than from yak's milk, is available in restaurants in this country, but it may take a visit to Tibet to really understand exactly why it is so widely consumed.

It took just one sip of Malaysian limeade for me to understand its appeal. In tropical countries, it is crucial to replenish your body's supply of salt, which is constantly being depleted through perspiration, which forms a film over our skin that helps keep us cool. A pinch of salt in a cool drink is an efficient way to replenish lost salt, and the beverage itself is a yummy combination of sweet, sour, and salty flavors. I make it at home all the time, especially in hot weather.

Tea-drinking habits in India were influenced by decades of the British Raj. Black tea sweetened with sugar and smoothed with milk is ubiquitous in both Great Britain and India, but Indians went a step beyond the English style. They added spices, including black pepper, to create chai, a sweet, milky black tea that has an enticing richness and depth. Oddly, chai has become extremely popular in the United States, where it is typically purchased, not made at home. I find this puzzling, as it is so easy to make it yourself and so much more delicious when you do.

We've yet to see salted coffee on supermarket shelves but I wouldn't be surprised if it showed up soon. Food Network Star Alton Brown has popularized what is being called "Man Coffee," which is simply freshly made coffee–preferably in a French press—with a pinch of kosher salt. If you drink dark roast, the salt will smooth out some of the bitterness so I do understand the appeal, though I prefer the flavor of medium roast coffee, preferably from Kona or another part of Hawaii Moku, which has almost no bitterness at all.

Since the first edition of this book, an artisan liquor, liqueur, and cocktail trend has exploded, with high-end bars filled with house-made infusions. More cocktails than ever are salted. Boutique distilleries offer hand-crafted liquors, including many infused with black pepper and a few with green peppercorns. Some bartenders even add a rasher of crisp bacon to everything from Bloody Marys and Martinis to their own signature cocktails. Have at it, mixologists!

Asian Limeade

Spiced Lassi

The Simplest Chai

My Favorite Chai

Mulled Hard Cider

Black Pepper Vodka

Sour Michelada

Salt-Rimmed Salty Dog, with Classic Variations

Peppery Mojito

Malaysian Martini

Spicy Margaritas

The Best Bloody Mary

Asian Limeade

Serves 1

In a tropical climate, people need to replenish the salt constantly lost through perspiration. One way they do it is to add a big pinch of salt to cooling drinks such as limeade and lemonade, a combination I find remarkably good (much better than Gatorade, our closest equivalent, which athletes use to replace salt lost through vigorous exertion). Neither the limes nor the lemons are quite as sour as ours are, nor is as much sugar added. Thus, the icy drink is subtle, mildly sweet, mildly sour, and mildly salty, perfect in the afternoon when you're feeling limp from heat and humidity. The first time I tried the real thing was in Kuching on the island of Borneo, while sitting in a little kiosk café near the Heroes' Memorial and Botanical Gardens. Geckos, the same pale green color as my limeade, scurried across the walls as I sipped and I don't think I've ever felt so far from home. Trying to capture the exact flavors I remember, I've experimented with different combinations; this easy, casual version comes closest. For complete authenticity, do not remove the seeds.

In India, you'll find a lemonade spiked with coarsely ground aromatic black salt (kala namak).

1 lemon or lime
2 teaspoons Simple Syrup (page 365, omitting the pepper, or from any basic recipe)
⅛ to ¼ teaspoon kosher salt
Ice cubes

Fill a large drinking glass three-quarters full with water. Cut the lemon or lime in half and squeeze the juice into the water. Add the Simple Syrup and salt, stir, and fill the glass to the top with ice cubes.

Spiced Lassi

Serves 1 or 2

Lassi is one of several drinks common in areas of the world where yogurt is eaten daily. In the United States, we see lassi most often in Indian restaurants (kefir, also made of yogurt, is more common in markets), but it is very easy to make at home. There are numerous sweet versions, but the savory one always includes a little salt, a little cumin, and frequently cayenne. Indian black salt contributes an aroma and flavor of sulfur; if you find it unpleasant, use another salt in its place.

1 cup plain whole milk yogurt, without additions such as agar, gelatin, or dry milk solids
1 teaspoon muddled fresh mint leaves or cilantro leaves
½ teaspoon crushed India black salt
¼ teaspoon cumin seed, toasted and crushed
Pinch of ground cayenne or other spicy ground chile
Ice

Whisk together the yogurt, herbs, salt, and ¾ cup water; add an additional ¼ cup water for a thinner drink. Stir in the cumin seed and pinch of cayenne. Fill glasses with ice cubes and pour the lassi over the ice. Sprinkle with the mint or cilantro leaves, and serve immediately.

The Simplest Chai

Serves 2

Chai has become remarkably popular in the 1990s and it is sold everywhere, already made and packaged in little boxes for retail sales and gallon jugs for sales to coffee houses and such. To me, it makes about as much sense to buy commercial chai as it does to buy tea that's been steeped and packaged. Why would you? There are many versions of the sweet and spicy drink, and I've been enjoying this one since I drank it daily during a two-month stay in India in the 1970s. In this simple version, India's black peppercorns play a primary role. Although all traditional chai includes black pepper, contemporary recipes sometimes omit it and add other ingredients such as ginger, mint, and nutmeg. Commercial versions are often loaded with cloves, which should be used sparingly, if at all.

2 cups water
1 cup milk
2 teaspoons loose black tea leaves, such as Assam
6 black peppercorns (Tellicherry), coarsely crushed
One 1-inch piece of cinnamon
2 whole cardamom pods
2 tablespoons sugar

Combine all of the ingredients in a small saucepan, set over medium heat, bring to a boil, remove from the heat, cover, and let steep for 10 minutes. Strain and serve.

My Favorite Chai

Makes 8 cups

As soon as winter rains begin or, in drought years, when temperatures plummet, I make chai and this has become the version I make most often, though I don't really follow a recipe. Sometimes I use ginger, sometimes I don't. Sometimes I use just black peppercorns, though I do enjoy the additional layer of flavor contributed by white peppercorns. Occasionally I forget the nutmeg. I do, quite frequently, double the recipe so that i have it on hand when friends and family visit. Everyone seems to love true chai, especially in cold weather. To double this recipe, increase everything except the ginger and cinnamon. Taste the chai after straining it and adjust the sugar to suit your palate. And no, you should not use nonfat, low fat, soy, or rice milk.

1 quart whole milk, preferably organic

½ cup sugar, plus more to taste

2 tablespoons black tea leaves (see Note below)

2 fresh ginger slices, gently crushed, optional

1 3-inch cinnamon stick

1 teaspoon black peppercorns

½ teaspoon white peppercorns, optional

2 whole cloves

4 green cardamom pods, crushed open

Several gratings whole nutmeg

Pour 4 cups of water into a large saucepan, add the milk and sugar and bring to a boil. Immediately lower the heat and stir in the tea. Add the ginger, cinnamon, peppercorns, cloves, cardamom, and nutmeg, stir, cover and remove from the heat. Let steep 10 to 12 minutes.

Return the pan to medium heat and just before the mixture reaches a boil, strain it into a teapot or individual cups. Store leftover tea in a glass jar or pitcher, covered, for 3 to 4 days. Reheat before serving.

NOTE

I prefer the delicacy of a good Darjeeling tea but sometimes enjoy the fuller flavor of Assam and so combine the two. English breakfast, Irish breakfast, and Queen Anne teas also make excellent chai.

Mulled Hard Cider

Serves 4

Hard cider, once the most common beverage in the United States—it was safer than water and drunk throughout the day, including by children—has made a strong comeback in the last few years. From coast to coast, artisan cider makers are crafting delicious beverages, some that rival wines, especially when it comes to pairing them at the table. Use an artisan hard cider here if you like or opt for non-alcoholic apple juice or cider if you prefer. From late fall through the coldest part of winter, hot cider, spiced or not, is a wonderful, warming beverage, excellent in the afternoon in place of tea or at night by a hot fire.

1 bottle (750 ml) hard cider
2 strips lemon zest
2 strips orange zest
Juice of 1 lemon
Juice of 1 orange
One 2-inch piece vanilla bean
1 stick cinnamon
3 or 4 whole cloves
Generous pinch of toasted, crushed, and sifted Sichuan peppercorns (see
    Note, page 271)
Black pepper in a mill
Kosher salt

Pour the cider into a large saucepan, add the lemon and orange zests, lemon juice, orange juice, vanilla bean, cinnamon, cloves, and Sichuan peppercorns. Add 6 or 7 turns of black pepper and a pinch of salt. Set over medium heat and just before it begins to simmer, remove it from the heat and cover the pan. Let steep 30 minutes, strain into a tea pot and serve immediately.

A Salt & Pepper Cookbook

Black Pepper Vodka

Makes 1 pint

Flavored vodkas have recently become extremely popular in the United States, but they have been made for decades in Russia and by people of Russian heritage. Black pepper is one of the most common flavorings; it adds a floral aroma and a spicy richness to the sharp, hot taste of vodka. In Poland, black pepper is added to vodka, the country's most popular alcoholic beverage, and drunk to relieve a stomachache. In Russia, the combination is used as a cure for the common cold.

Why make your own, you may wonder, now that commercial versions are available. Why not? It's easy, you have control over all the ingredients and you'll have the satisfaction of doing it yourself.

1 tablespoon black peppercorns or 2 tablespoons mixed peppercorns
16 ounces vodka

Combine the peppercorns and vodka in a glass jar or bottle. Store in a cool, dark cupboard for 3 weeks. Store in the freezer and serve over ice or in mixed drinks, such as Bloody Marys

Sour Michelada

Makes 1

A michelada can be as simple as a cold beer with a splash of hot sauce and as complicated as you want to make it, with fresh tomato juice, a pantry of spices and herbs, and more. This is my favorite version, which I prefer with one of the newly-available sour pilsners. When none are available, I make my micheladas with Bohemia, my favorite South-of-the-Border beer.

1 lime wedge
Chipotle Salt (page 342), in a saucer
Crushed ice
Juice of 1 lime
1 tablespoon muddled cilantro (see Note below)
1 bottle sour beer of choice, well chilled
Cholula hot sauce, Crystal hot sauce or other hot sauce of choice

Rub the rim or a pint beer glass with the lime wedge and then gently turn the rim in the salt.

Fill the glass half full with ice, pour in the lime juice, add the muddled cilantro, tilt the glass and carefully pour in the beer. Add hot sauce to taste and enjoy.

NOTE

"Muddled," when referring to herbs, simply means smashed or crushed; it is a term typically used by bartenders, who have special muddling devises, made of stone, wood, bamboo, or stainless steel. For this drink, put 6 to 8 cilantro leaves into a small bowl or mortar, add a pinch of kosher salt and pound the leaves a few times.

Salt-Rimmed Salty Dog, with Classic Variations

When I was writing Vinaigrettes and Other Dressing *(Harvard Common Press, 2013, $16.95), I fell in love with the combination of fresh grapefruit juice and fragrant ground cumin. It is delicious in this context, as well as in a simple grapefruit vinaigrette.*

Flake salt, on a plate
2 generous pinches of ground cardamom
Small grapefruit wedge
Crushed ice or ice cubes
3 to 4 ounces vodka
16 ounces grapefruit juice
½ teaspoon kosher salt

Mix the salt with one pinch of cardamom.

Moisten the rim of two highball glasses (medium, i.e., 8 to 12 ounce, tumblers) with the grapefruit wedge and then dip the rim into the salt.

Fill the glasses nearly full with ice.

Pour the vodka and grapefruit juice into a cocktail shaker, add the ½ teaspoon of salt and remaining pinch of cardamom, stir with a long spoon until the salt is dissolved and pour over the ice.

Enjoy immediately.

Variations:

- Salty John: Replace the vodka with whiskey.

- Salty Juan: Replace the grapefruit wedge with a lime wedge, the vodka with tequila, and the cardamom with chipotle powder.

- Salt Lick: Use lemon wedges instead of grapefruit wedges to moisten the rim of the glasses and omit the cardamom. Mix together 3 ounces vodka, 4 ounces bitter lemon soda, and 4 ounces grapefruit juice and pour over the ice.

A Salt & Pepper Cookbook

Peppery Mojito

Serves 1

    A mojito is one of the most refreshing cocktails there is during a heatwave. I do not add club soda, as I find ice provides the perfect amount of dilution.

6 to 8 fresh spearmint leaves
5 or 6 fresh cilantro leaves
Kosher salt
2 thin slices of lime, seeded and cut in half
½ teaspoon freshly grated ginger
2 tablespoons Simple Syrup with Pepper, plus more to taste
Crushed ice
1 jigger (1 ½ ounces) white rum of choice
Juice of 1 lime

Put the mint and cilantro into a chilled pint glass, add a pinch of salt and lightly crush the leaves, using a muddler or a narrow wooden pestle. Add the lime, ginger, and simple syrup and pound gently a few times to lightly crush the lime. Fill the glass ½ to ⅔ full with ice, add the rum and the lime juice and stir will with a long spoon. Taste and add a bit more simple syrup if needed for balance.

Malaysian Martini

Serves 1

The lounge revival of the mid 1990s saw the return of such retro pleasures as Martin Denny music, the little black dress, and the martini. College students hosted cocktail parties rather than beer bashes, and martini bars were set up everywhere from art openings to corporate galas. Bars specializing in martinis opened in several cities, and some of their concoctions stretched the imagination of even the trendiest imbiber, with chocolate martinis, bacon martinis, and martinis that tasted a lot like the liquid penicillin I remember from childhood. This version is closer to the original classic martini and is not as trendy as you might think. Vodka has been flavored with peppercorns for centuries in Russia. It might be more fitting to include the traditional gin, particularly because gin was often the drink of choice among colonists in Malaysia, where today peppercorns are farmed, but I prefer the flavor of vodka and pepper to that of gin and pepper. If I were making this on the island of Borneo, near one of the thousands of peppercorn farms that cover the island, I would pinch off a cluster of fresh peppercorns as a garnish for each drink.

Cracked ice cubes
3 ounces Black Pepper Vodka (page 335)
½ ounce sweet Italian vermouth
2 dashes orange bitters

Fill a cocktail shaker with cracked ice, add the vodka, vermouth, and bitters, and shake for several seconds, until the shaker is frosty. Strain into a chilled stemmed glass and serve immediately.

A Salt & Pepper Cookbook

Spicy Margaritas

Serves 2

    Dried chipotles add another dimension to the already smoky flavor of tequila. Put two chipotles in a bottle with 8 ounces or so of tequila and let the mixture steep for a few weeks. The result is a rusty gold liquid with a heady tequila aroma mingled with the irresistible scent of smoked chiles. It is delicious—strong, spicy, and bold. Be advised, these margaritas are not for the timid!

3 lime wedges
Chipotle Salt (page 342) in a saucer
Cracked ice cubes
2 ½ ounces chipotle tequila, or other tequila of choice
⅓ cup fresh lime juice
1 ounce Cointreau

Rub the rim of each of two glasses with one of the lime wedges and then dip the rim of each glass into the saucer of Chipotle Salt. Fill the glasses with cracked ice cubes, add half the tequila to each glass, followed by half the lime juice and half the Cointreau. Stir with a cocktail spoon, garnish with a lime wedge, and serve immediately.

The Best Bloody Mary

Serves 1

My favorite time to enjoy a Bloody Mary is at an airport, especially a small one, when I can look out the window of a tiny bar and watch the planes come and go. At home, I make the drink only when I have fresh summer tomatoes.

1 teaspoon celery salt
Flake salt on a plate
Wedge of lemon
Crushed ice or ice cubes
3 ounces (2 jiggers) Black Pepper Vodka (page 335) or other vodka of choice
Generous squeeze of fresh lemon juice
2 or 3 shakes of Worcestershire sauce
3 or 4 drops of Tabasco sauce
6 to 8 ounces very fresh tomato juice (see Note on next page)
Sel gris, ground
Black pepper in a mill
Small celery stalk

Add the celery salt to the flake salt, rub the rim of a large highball glass with the lemon and invert it onto the flake to coat it with the salt mixture.

Fill the glass with ice cubes.

Pour in the vodka and lemon juice, add the Worcestershire sauce and Tabasco sauce and carefully add the tomato juice, stopping before it touches the salt rim.

Add a pinch or two of sel gris, a few turns of black pepper and stir very gently with a long spoon.

Add the celery stalk and serve.

NOTE

The very best tomato juice to use in this drink is that which drains away when you make tomato concasse, which is simply tomatoes that have been peeled, seeded, minced and left to drain in a strainer set over a deep bowl. Some chefs call this liquid tomato water. Whatever you call it, it is absolutely lovely and bursting with true tomato flavor (if you use ripe, in-season tomatoes, of course.)

Variations:

- For a Virgin Mary, simply omit the vodka and add more juice.

- For a Bloody Maria, use chipotle salt in place of the celery salt, a lime wedge and lime juice instead of lemon, and replace the vodka with your tequila of choice.

Sauces, Preserves, Condiments

In this chapter, I focus on some favorite combinations of flavors that typically require other foods to complete them. One doesn't serve, say, a mignonette without an oyster and one rarely takes a swig of vinaigrette or a spoonful of chutney. That said, the divisions are somewhat amorphous, permeable. Several of the recipes in the appetizer chapter could fit here, especially hummus and skordalia. They are in the appetizer section because they are most often served neat, with just crackers or bread, rather than as part of a larger recipe.

The reverse is true of recipes I place here. As good as, say, artichoke and olive tapenade may be by a spoonful, it needs something to complete it.

Gomashio

Peppery Mignonette

Preserved Lemon Chermoula

Preserved Lemon Gremolata

Artichoke & Olive Tapenade

Spicy Salt Egg Sauce

Spicy Maple Butter

Black Pepper Dressing

Green Peppercorn Mustard

Green Peppercorn Mayonnaise

Bacon Vinaigrette

Sarawak Sambal

Raisin, Onion & Green Peppercorn Chutney

Hawaiian Chile Water

Peppercorn Oil

Simple Syrup with Pepper

Gomashio

Makes about ½ cup

Gomashio is a Japanese condiment, a type of furikake, traditionally used to season rice, There are many commercial brands available in Asian markets.

Making your own is easy and inexpensive. In December, I make several pounds of it, put it into small canning jars, and donate it to my hula school's annual holiday craft fair, where it always sells out. I also give it as a gift, too, and have occasionally sold a few jars when I've done book signings at local farmers markets. When I make it in large batches, I typically grind the sesame seeds in a food processor fitted with its metal blade.

Because I live near the Pacific Ocean, I use local seaweed. Sometimes I use Portuguese sea salt, sometimes Hawaiian alaea salt, and sometimes sel gris. It is important that whatever salt you use is quite dry.

½ cup organic white sesame seeds
1 tablespoon unrefined solar dried sea salt
Small handful nori or wakame
½ teaspoon red pepper flakes, optional

Set a small heavy sauté pan over medium-low heat, add the sesame seeds and toast, stirring all the while, until they are golden brown and fragrant. Do not let them burn.

Immediately transfer the sesame seeds to a suribachi or mortar and return the pan to the heat. If the salt has any moisture in it, put it in the pan and stir until it turns opaque and is completely dry. Cool and add the salt to the suribachi or mortar.

With the pan over high heat, toast the seaweed until it is just crisp.

Let it cool, break it into pieces and add it to the suribachi.

Using a wooden pestle, grind the mixture together until most of the seeds are broken but not reduced to a powder. Add the red pepper flakes, if using, and mix well.

Store in a glass jar and use as needed.

Peppery Mignonette

Makes about ½ cup

The word "mignonette" refers to coarsely cracked white peppercorns. It also is the name of the classic condiment served with raw oysters on the half shell, and that name likely refers to the pepper in the sauce. There are many versions, made with sherry vinegar, balsamic vinegar, raspberry vinegar, or rice vinegar seasoned with ginger. If the vinegar is good, the sauce will be good, though those with too aggressive a taste (sherry and balsamic vinegars, in my opinion) overpower the delicate sea flavors of an oyster, which seems to concentrate the ocean's salty savor better than any other creature does. This recipe makes enough sauce for about four dozen small to medium oysters; each one needs just a scant half tea-spoon of mignonette, enough for a tiny burst of peppery tartness to mingle with an oyster's natural saltiness.

½ teaspoon flake salt or ¾ teaspoon sel gris
½ cup best–quality white wine vinegar, such as Vinaigre de Banyuls
1 finely minced shallot (about 2 tablespoons)
½ teaspoon crushed white peppercorns
1 teaspoon crushed black peppercorns
2 tablespoons fresh lemon juice

Stir the salt into the vinegar and when it is dissolved, add the shallot, peppercorns, and lemon juice. Chill for 30 minutes before serving. Mignonette may be stored in the refrigerator, covered, for a week or so.

Preserved Lemon Chermoula

Makes about ¾ cup

Chermoula is a staple in my kitchen, like ketchup, say, or steak sauce is in others. I enjoy it on yogurt, scrambled eggs, omelets, potato soup, all manner of stews, and almost any type of fish. This version calls for preserved lemons. If you have a sudden craving for it but don't have any preserved lemons in the pantry, use the juice of a whole lemon, plus a bit more to taste.

3 or 4 garlic cloves, peeled
Flake salt
½ cup lightly packed fresh cilantro, chopped
¼ cup lightly packed fresh Italian parsley, chopped
1 teaspoon sweet paprika, preferably Spanish
1 teaspoon hot paprika, preferably Spanish
½ teaspoon chipotle powder, piment d'Esplette or other hot ground chile
½ teaspoon ground cumin
3 preserved lemon wedges, minced (page 281)
Juice of ½ lemon, or to taste
Black pepper in a mill
¼ to ⅓ cup extra-virgin olive oil, plus more to taste

Put the garlic in a suribachi or mortar, sprinkle with salt and use a wooden pestle to crush the garlic into a paste. Add the cilantro and parsley and continue to grind with the wooden pestle until a uniform but chunky purée is formed. Add the paprikas, ground chile, cumin, and preserved lemons and stir in the lemon juice.

Season with salt and several turns of black pepper and stir in the olive oil. Taste and correct for salt and acid as needed. Cover and chill; remove from the refrigerator 30 minutes before using.

Chermoula will keep up to 2 days in the refrigerator, but it is best the day it is made.

A Salt & Pepper Cookbook

Preserved Lemon Gremolata

Makes about ½ cup

Traditional gremolata, a classic condiment with osso buco, is a simple blend of lemon zest, garlic and parsley. I've never found it is good as it should be, as without salt it just doesn't come alive. In this version, salt is contributed by the preserved lemons and chopped cilantro adds yet another layer of flavor. Use with grilled or roasted fish or chicken, with meat stews and over creamy polenta.

4 or 5 preserved lemon wedges (page 281), seeded and chopped
¾ cup, lightly packed, Italian parsley leaves, chopped
¼ cup, lightly packed, cilantro leaves, chopped
3 to 4 garlic cloves, minced

Put the chopped lemon, parsley, cilantro, and garlic into a small bowl and toss together gently with a fork. Cover and set aside until ready to use. Gremolata is best used right away.

Artichoke & Olive Tapenade

Makes about 2 ¼ cups

Tapenade is one of the classic salty condiments of southern Europe, made in one version or another virtually everywhere olives grow. It might be a smooth purée, it might be chunky, but it almost always includes brine-cured olives, garlic, and anchovies. In this version, I've added two other ingredients common to Provence and northern Italy, artichokes and walnuts, and combined them with green peppercorns for a chunky sauce that is excellent on crostini, with cheese, or tossed with hot pasta.

1 tablespoon kosher salt
3 large artichokes
1 tablespoon olive oil
3 garlic cloves
2 or 3 anchovy fillets, drained
2 teaspoons green peppercorns in brine, drained
1 teaspoon minced lemon zest
1 tablespoon fresh lemon juice
½ cup extra virgin olive oil
1 cup (6 ounces) cracked green olives, such as picholine, pitted and minced
¼ cup (2 ounces) walnut pieces, toasted and minced
1 tablespoon flat-leaf parsley, minced
Black pepper in a mill

Fill a large pot two-thirds full with water, add the salt, and bring to a boil over high heat. Using kitchen shears, snip off the tips of the outer leaves of the artichokes. Drizzle a little of the olive oil in the center of each artichoke and place them in the boiling water. Return to a boil, cover, reduce the heat, and simmer until the artichokes are tender, from 20 to 40 minutes, depending on their size and age.

Meanwhile, use a suribachi or mortar and pestle to grind the garlic and anchovies together until they form a smooth paste. Fold in the green peppercorns, lemon zest and juice, and olive oil, scraping the sides of the bowl as you mix. Set aside.

Transfer the artichokes to a colander or strainer, rinse them under cool water, drain thoroughly, and let them cool. When they are cool enough to handle, remove all of the leaves, reserving them for another use. Use a teaspoon or a small, sharp paring knife to cut away the choke in the center of each artichoke heart. Discard the chokes and cut the hearts into ¼-inch dice.

In a medium bowl, combine the diced artichoke hearts, olives, walnuts, and olive oil mixture. Add the parsley and several turns of black pepper. Taste and season with salt. Let the tapenade rest 30 minutes before serving.

Spicy Salt Egg Sauce

Makes about ½ cup

This sauce, Thai in origin, is so good, one of those enticing combinations of flavors that calls out to you from the refrigerator at odd hours. Be careful, or you won't have any actually to serve. Although the sauce is most intense when made with salted eggs, you can make a good version with fresh eggs; Just add more salt. The same is true with dried shrimp paste; it is a traditional ingredient but if you don't have it, simply omit it and add another splash of fish sauce.

I use Salt Egg Sauce on baked and grilled fish; it is also an excellent condiment with simple steamed rice, Salted Mustard Greens (page 279), and sliced summer tomatoes.

4 Salt Eggs (page 275) or 4 fresh eggs, hard-cooked and cooled
3 cilantro roots, washed and trimmed
4 garlic doves, crushed
2 serrano chiles, chopped
2 shallots, chopped
1 teaspoon dried shrimp paste, optional
2 teaspoons sugar
2 tablespoons fish sauce
Juice of 2 limes
2 scallions, white and green part, trimmed and minced

Peel the eggs, separate the whites from the yolks, and set the whites aside for another use. Press the yolks through a sieve, grate them on the small blade of a grater, or mash them with a fork and put them into a small bowl.

Using a mortar and pestle or a suribachi, grind the cilantro roots and garlic until they form a paste. Add the serranos and shallots and continue to grind until they form a uniform and almost smooth paste. Add the shrimp paste, if using, and sugar, mix thoroughly, and then use a rubber spatula to fold in the egg yolks.

Stir in the fish sauce and lime juice, transfer to a small serving bowl and fold in the scallions. Cover the sauce and let it rest at room temperature for 30 minutes before serving.

Spicy Maple Butter

Makes about ⅓ cup

There is a problem with this sauce. It's too good. If you taste it to correct for salt, you are tempted to simply drink it up in one long glorious velvety swig.

3 tablespoons organic butter
2 tablespoons best-quality maple syrup
1 tablespoon lemon juice
Black pepper in a mill
¾ teaspoon ground cardamom
¼ teaspoon ground clove
Pinch of flake salt

Put the butter into a small saucepan set over medium heat, add the maple syrup, the lemon juice and several very generous turns of pepper. Add the cardamom, clove, and a generous pinch of salt. Swirl and remove from the heat.

Serve hot.

Black Pepper Dressing

Makes about ⅔ cup

This sauce takes just a couple of minutes to make and it is both delicious and versatile. Spoon it over avocado and cottage cheese for a quick lunch, use it as a dip for steamed artichokes, broiled shrimp, and vegetable crudités, and as a sauce for grilled fish. It is also excellent on a wide range of salads, including Avocado, Grapefruit, & Chicken Salad. on page 143.

2 limes
2 teaspoons kosher salt
3 tablespoons black peppercorns, coarsely crushed
⅔ cup extra virgin olive oil

Peel the zest off one of the limes, mince it, and set it aside. Juice the two limes and combine the lime juice (about 3 tablespoons) and salt in a small bowl, stirring until the salt is dissolved. Add the peppercorns, lime zest, and olive oil, and set the dressing aside, covered, until ready to use.

Variation:

Use lemon juice and lemon zest in place of the lime.

Green Peppercorn Mustard

Makes about ⅓ cup

I always use PIC Dijon (see resources, page 394) when I make flavored mustards. It's perfectly balanced, and fairly inexpensive. In its place, you can use Grey Poupon, although you might want to add 1 tablespoon mustard flour (such as Colman's) mixed with 2 teaspoons cold water to give it a little extra heat. This mustard, if refrigerated, will maintain peak flavor for about two weeks. Use it on sandwiches, with meatloaf, and with pâtés and meat terrines.

6 tablespoons Dijon mustard
2 teaspoons crushed dried green peppercorns
1 tablespoon whole dried green peppercorns

Combine the ingredients in a small bowl or glass jar, cover, and refrigerate overnight before using.

Green Peppercorn Mayonnaise

Makes ½ cup

This mayonnaise is excellent with gravlax (pages 268 and 270), with radishes, and on many sandwiches, from roasted eggplant and salami to BLTs and grilled salmon on focaccia.

½ cup mayonnaise, homemade or Best Foods/Hellman's brand
1 ½ to 2 teaspoons crushed dried green peppercorns
1 teaspoon fresh minced lemon zest

In a small bowl, combine the mayonnaise with the peppercorns and lemon zest. Cover, refrigerate, and let rest at least 30 minutes before serving. Store, covered, in the refrigerator.

Variation:

In place of the dried peppercorns, use 2 teaspoons green peppercorns in brine, strained and crushed. Omit the lemon zest and add ¼ teaspoon each ground white pepper and ground black pepper.

Bacon Vinaigrette

Makes about ¾ cup

A warm vinaigrette is one of the most diverse sauces in any cook's repertoire. It can be used on certain traditional salads—German potato salad and warm spinach salad, for example—but is also delicious spooned over an omelet, roasted sweet potatoes, grilled fish, grain salads, and more. Here, layers of salt and pepper build in delicious flavors. For a final flourish, finish the dish you are cooking with a light sprinkling of smoked salt.

4 slices bacon
2 medium shallots, minced
2 garlic cloves, minced
Pinch of kosher salt
½ cup best-quality white wine vinegar
1 tablespoon fresh lemon juice
Black pepper in a mill
½ cup best-quality extra virgin olive oil
3 tablespoons chopped fresh Italian parsley

Fry the bacon in a heavy sauté pan until it is crisp; transfer to absorbent paper to drain.

Add the shallots to the bacon fat and sauté until soft, 6 to 7 minutes; add the garlic and sauté 2 minutes more.

Add the vinegar and simmer until it is reduced by half. Add the lemon juice, season generously with black pepper, add the olive oil, heat through and turn off the burner. Taste, correct for salt and acid balance and stir in the parsley. Chop or crumble the bacon, add it the vinaigrette and use while hot.

A Salt & Pepper Cookbook

Sarawak Sambal

Makes about ⅔ cup

A sambal is a savory condiment, typically but not always fairly spicy, common in India and Southeast Asia. I first tasted this condiment in the test kitchen of the Pepper Marketing Board in Kuching, the capital of Sarawak, Malaysia. Moments before, I had been handed a jar of Creamy White Pepper, which fell out of my hands as I removed the lid to smell the pepper. Glass shattered, peppercorns darted all over the floor of the immaculate laboratory, I was mortified, and my charming host quickly shifted our attention to a selection of condiments that Shazat Khan, the PMB's chef, had developed. This was my favorite.

Dried anchovies are common throughout Asia and they are very different from the canned anchovies we are used to in the West. Look for them in Asian markets, as you will not get the proper flavor if you used canned or even salted ones.

1 cup (1 ½ ounces) dried anchovies
4 large shallots
2 teaspoons kosher salt
5 garlic cloves
¼ cup (1 ½ ounces) green peppercorns in brine, drained
⅓ cup mild olive oil, or other mild vegetable oil
1 teaspoon sugar
1 tablespoon rice vinegar

Soak the anchovies in cool water for at least 30 minutes, drain, and set them aside on a tea towel to dry. Using a suribachi or mortar and pestle, crush the shallots and 1 teaspoon of the salt together, add the garlic and peppercorns, and grind to a paste. Heat the oil in a medium sauté pan and fry the anchovies until they are almost crisp; remove them from the pan with a slotted spoon and set them aside. Sauté the shallot mixture in the same oil, stirring continuously, until it is very fragrant and beginning to brown; do not allow it to burn. Stir in the remaining teaspoon of salt, the sugar, and the vinegar, remove from the heat, and add the anchovies. Serve immediately, or refrigerate, covered, for up to 5 days.

Raisin, Onion & Green Peppercorn Chutney

Makes about 4 cups

A chutney is an ideal context in which to highlight the tangy flavor of green peppercorns preserved in brine; their flavor merges perfectly with the vinegar in the chutney and the peppercorns themselves provide spicy bursts of flavor in this hot-sweet condiment. Serve this chutney as you would any traditional cooked chutney, with curries, biryani, and tandoori meats. It is also excellent with American-style grilled and roasted meats and poultry.

1 pound raisins
1 ½ cups (12 ounces) sugar
2 ½ cups apple cider vinegar
2 medium yellow onions, halved and thinly sliced
8 large garlic cloves, minced
One 2-inch piece ginger (about 1 inch in diameter), peeled and cut into very small julienne
2 tablespoons green peppercorns in brine
1 teaspoon red pepper flakes
1 teaspoon kosher salt
One 3-inch piece of cinnamon
3 whole cardamom pods
3 whole cloves

In a large nonreactive pot, combine the raisins, sugar, vinegar, onion, and garlic, and set over medium heat. Bring to a boil and stir continuously until the sugar is dissolved. Add the ginger, peppercorns, red pepper, salt, cinnamon, cardamom, and cloves, reduce the heat to low, cover, and simmer until the raisins are tender and the chutney very thick, about 2 hours.

Ladle into hot clean jars and store in the refrigerator for up to 3 months, or ladle into sterilized jars and process in a hot-water bath (see Note about canning, page 273). Cool, check seals, and store in a cool dark cupboard. Refrigerate after opening.

Canned chutneys typically last about a year before opening.

Hawaiian Chile Water

Makes 2 cups

Hawaiian food is typically mild, though Chili Water, often offered alongside, allows you to spike your kalua pig, white rice, poi, ribs, or any other savory foods with a bit of heat. You don't see it in high-end restaurants in Hawaii but home-style cafés, those that serve plate lunch, typically have it on the table next to soy sauce, salt, and pepper. I have fallen in love with it and use it on many dishes, not just Hawaiian ones.

2 teaspoons Hawaiian alaea salt
2 teaspoons white vinegar
1 garlic clove, crushed
6 to 8 small hot chili peppers, such as Thai

Put a 16-ounce glass bottle into boiling water and simmer for 10 minutes. Remove from the heat, cool slightly and use tongs to transfer to a work surface.

Put the salt, vinegar, and chiles into the bottle and fill with tap water, spring water, or filtered water. Add the bottle's lid, shake and set aside for 2 days so that flavors blossom. This keeps well in the refrigerator for several weeks and can be topped off as it is used.

Use as a condiment with kalua pig, roasted sweet potatoes, poi, rice, and with any other dish where you want a splash of salty heat.

Peppercorn Oil

Makes 3 ½ cups

The volatile oils of the peppercorns infuse olive oil with a lusty heat and aroma; use this oil to make dressings and as a condiment where you want the flavor but not the texture of pepper. A buttery Ligurian olive oil or a California Mission olive oil is ideal for this recipe.

¼ cup (1 ounce) white peppercorns
½ cup (2 ounces) black peppercorns
¼ cup (½ ounce) dried green peppercorns
3 cups extra virgin olive oil

Toast the Sichuan peppercorns in a small dry sauté pan set over medium-low heat until they are fragrant, about 3 to 4 minutes. Remove them from the heat and let them cool. Put all the peppercorns and the coriander in a glass quart jar, pour the olive oil over the peppercorns, seal the jar with its lid, and place in a cool, dark cupboard for 2 to 3 weeks. Strain, store in a clean glass jar or bottle, and use within 3 months. Use the peppercorns to make a second batch of oil, letting it sit for 3 to 4 weeks before straining. Discard the peppercorns after making the second batch of oil.

> **Variation:**
>
> For a more fragrant oil, toast ¼ cup Sichuan peppercorns in a small dry pan set over medium heat until they are fragrant, about 3 to 4 minutes. Cool, crush, sift and discard the hard seed cores. Add the delicate outer part of the Sichuan pepper to the other peppercorns, along with ¼ cup coriander seeds and continue as directed in the main recipe.

Simple Syrup with Pepper

Makes about 2 cups

Simple syrup, sometimes called bar syrup or bar sugar, is simply granulated sugar simmered with water until it is clear. Other ingredients—peppercorns, other spices, fresh ginger—can be added, too. What I like about this version is that the flavors of the peppercorns blossom more fully than when pepper is added to a dish at the last minute.

2 cups sugar
1 teaspoon cracked white peppercorns
2 teaspoons cracked black peppercorns

Combine the sugar, white peppercorns, and 1 cup water in a small heavy saucepan set over high heat; do not stir. Bring to a boil, reduce the heat, and simmer for 4 to 5 minutes, until the sugar is completely dissolved and the syrup transparent. Remove from the heat, add the black peppercorns, cover, and cool to room temperature. Cover and let rest overnight. Strain into a glass jar, cover, refrigerate, and use as needed.

Seasoned Salts & Other Spice Blends

One hot afternoon in Kuching, Sarawak, I walked beyond the waterfront and its familiar food stalls and tourist shops to a string of local markets selling meat, poultry, and mounds of familiar and unfamiliar fruits and vegetables. I was hoping to stumble across a few out-of-season mangosteens—the Queen of Fruit it is called, and rightly so: It is exquisite. What I found instead was a spice merchant. Certainly, I had seen vendors selling spices in markets around the world, but none sold them in quite this form. Instead of huge sacks of dried whole or ground spices, this merchant was selling spices that had been mixed into very thick pastes. Bright red chiles, coppery cumin, creamy cardamom, and orange turmeric had each been mixed with water and salt. Customers told the merchant what they wanted to cook—a chicken or fish curry, perhaps, or Sarawak laksa—and he sliced off just enough for the dish, wrapping the colorful slabs on top of one another in a single pandan leaf. He allowed me to taste them; each spice, its flavors lifted by salt, blossomed full and warm in my mouth.

This is, I believe, a unique moment in American cooking. Over the past two decades, we have grown comfortable with ingredients that previously were familiar only to relatively few dedicated individuals. In the process, the quality of our ingredients has improved immensely. We are no longer satisfied with spices and dried herbs that sit on grocers' shelves (or in our own pantries) for months, or even years. We know their flavors and we want them, full and bold, as they should be.

Basic Seasoned Salt

Chipotle Salt

Roasted Sichuan Pepper Salt

Caraway Salt

Cajun Spice Mixture

Mixed Peppercorns

Seasoned Salts

One morning I awoke to find a small package on my front porch. Dropped off by my friend Robert Kourik, a gardener and the author of lovely gardening and landscaping books, the package was filled with small pouches of seasoned sea salt from South Africa, the salt itself made by a tribe in Namibia. The pouches were attached to one another, a daisy chain of salty flavors: salt with Indian spices, with Asian spices, with celery salt, with crushed chiles, with herbs de Provence, and, of course, sea salt with crushed black pepper.

Another day, I received a tiny round glass jar filled with beautiful transparent salt crystals layered with green, black, and pink peppercorns, carried home from Italy by my friend Jerry Hertz. (The whole peppercorns and coarse salt don't exchange flavors, but the visual effect is striking.) The most unusual salt to come my way was a smoked salt from the north coast of Denmark, sent to me by Chef Lars Kronmark of the Culinary Institute of America in St. Helena, California. At that time, there was no smoked salts on the market in the United States. The salt arrived in a tightly sealed package a few days after a local newspaper ran my photograph with a story about salt. The translucent brown crystals of smoky-tasting salt were stunningly fragrant, but too much so, as the aroma permeated everything, including layers of sealed plastic. It was so intense that anyone who didn't know the source of the aroma would think there was a nearby fire. Since that time, several smoked salts have appeared on the market; most have more subtle aromas that won't get you suspected of arson. I find these salts, including that first one, quite delicious, especially with eggs, grilled fish, and delicate cheese soufflés.

Salt seasoned with dried herbs, ground spices, crushed chiles, lemon or lime juice, or a host of other ingredients is a simple way to add flavor to a finished dish. In making flavored salts, seeds should be toasted and ground, herbs should be dried and crumbled, and chiles should be ground before being combined with the salt. For the best flavor, make the salts in amounts you can use within a week or less if you are adding perishable ingredients, such as fresh juice.

Today, as I write the revised edition of this book, smoked salt, lavender salt, rosemary salt, and scores of other flavored salts are available in the marketplace. Some are good, some are a tad gymnastic for my taste, and some seem a bit silly. In the end, you get a very similar result simply applying salt to the ingredients when you cook them. That said, seasoned salts—beet salt, for example, and dried tomato salt—can be dazzling to the eye.

About Curry Powder and Other Spice Mixtures

Fragrant, spicy, floral, sour, and sweet—these are the elements that shape not only a good Indian curry powder but also similar spice mixtures used the world around. Specific ingredients vary among countries, regions, recipes, and cooks, but there are several ingredients that nearly always play a role, including black and white peppercorns, and often, salt.

Turmeric, coriander, cumin, fenugreek, ginger, cloves, cinnamon, black peppercorns, and varying amounts of cayenne pepper are found in most Indian curry mixtures. Depending on the type—sweet or hot, from Madras or Goa, for vegetable stews or roasted meats—amounts of these and other ingredients will vary a great deal. White or brown mustard seed, nutmeg, fennel, paprika, and cardamom might appear. If you are new to curry powders, you will not want to make your own—it's not difficult, but it can be time consuming to locate high-quality spices and prepare them properly. You must also use them fairly quickly, or they will lose their aroma and potency. To become acquainted with curry powder, purchase commercial mixtures from small companies (see resources, page 394) that ensure freshness.

Thai curries come in the form of pastes, with crushed fresh garlic, shallots, chiles, and fresh herbs. They, too, can be purchased, though if you fancy Thai cooking, you should take the time to make your own; they are superior to purchased types.

Classic African spice mixtures resemble, but are not identical to, Indian curries. One of the most classic European mixtures is quatre épìces, made with white pepper, nutmeg, cloves, and ginger.

In the United States, too, there are several spice mixtures that shape our regional cuisines. The classic Cajun spice mixture used in gumbos, jambalayas, beans, and rice dishes relies on onion powder, garlic powder, paprika, finely ground black and white pepper, ground cumin, mustard flour, dried thyme, oregano, basil, cayenne, and salt. There are countless variations of this mixture, many of them rubs for barbecue and the typical New Orleans dish, blackened redfish.

Crab boil, with black pepper, mustard seed, dill seed, coriander, cloves, allspice, ginger, and bay leaf, is common from Maryland south to Louisiana. In the South, a mixture of black pepper, nutmeg, coriander, cumin seed, cloves, cinnamon, green cardamom seeds, paprika, and ground chiles is used as a rub for meats and vegetables to be grilled.

The most important element in all of these mixtures is freshness. The flavors of spices can lose their intensity fairly quickly once they are ground, so it is crucial to buy from a reliable source, store the spices properly in a cool, dark cupboard, and use them or replace them within about six months.

Basic Seasoned Salt

Makes about ¼ cup

The reason to make seasoned salt at home is because there is a particular combination you really like. It is very easy to do. Many seasoned salts call for garlic powder and onion powder but, with a single exception, I prefer to use fresh versions of those ingredients. In Cajun cooking, you won't get traditional flavors without garlic and onion powder; you'll find a recipe for Cajun spice mixture on page 358.

You should use flake salt to make seasoned salts, as the uneven crystals will hold the ingredients in an even suspension. Hard salts, especially table salt, will precipitate out of the mixture and fall to the bottom of the container.

3 tablespoons flake salt, preferably Diamond Crystal Kosher Salt
1 to 3 teaspoons spices and/or dried herbs of choice

Put the salt into a small bowl.

If using whole seeds or leaves, crushed gently in a suribachi, add to the kosher salt and stir to combine. Store in a glass jar with a non-metallic closure and use within a few weeks.

Suggestion combinations:

- 2 teaspoons Chinese Five Spice Powder
- 2 allspice berries, 2 cloves, ½ star anise, crushed
- 2 teaspoons cumin seed, lightly toasted and crushed
- 2 teaspoons culinary grade lavender, lightly crushed, 2 teaspoons granulated sugar
- 1 teaspoon each sweet Spanish paprika, smoked Spanish paprika, and hot Spanish paprika
- 1 teaspoon fennel seed, lightly toasted, crushed, 1 teaspoon fennel pollen
- 1 teaspoon crushed dried thyme, 1 teaspoon crushed dried summer savory
- 2 teaspoons ground dried ginger, 1 teaspoon ground black peppercorns

Chipotle Salt

Makes about ½ cup

The best chipotle powders are made by small producers, many of them located in Sonoma County, California, where Tierra Vegetables, a farm located in Santa Rosa, launched a little industry when they first offered their chipotle powder in the 1980s. Now even the large spice manufacturers make it, though those big commercial versions are rather one dimensional.

Use chipotle salt on the rims of margarita glasses, on roasted poultry and grilled meats, and to finish bean dishes.

½ cup kosher salt
1 teaspoon chipotle powder, more or less to taste

Place the salt and chipotle powder in a small container with a lid, seal the container, and shake vigorously for several seconds until the chipotle powder is evenly distributed in the salt. You should not see large patches of red. Chipotle salt will keep indefinitely.

Roasted Sichuan Pepper Salt

Makes about 3 tablespoons

Although this aromatic salt has many more traditional uses—with chicken livers, potted meats, and pâtés, for example—I find it delicious on popcorn.

1 ½ teaspoons Sichuan peppercorns, coarsely crushed and sifted to remove seeds
3 tablespoons kosher salt

Place the peppercorns in a small heavy skillet set over medium heat and toast them, agitating the skillet slightly, until they start to give off an aroma, 3 or 4 minutes. Reduce the heat to low and slowly stir in the salt. Remove from the heat and let cool. Crush the mixture using a mortar and pestle, suribachi, or electric spice mill. Store in an airtight jar.

Caraway Salt

Makes about ¼ cup

In her book, The Cooking of the Eastern Mediterranean, *Paula Wolfert tells readers about discovering a flavored salt from the Svaneti region of Georgia, explaining that it is served as a condiment with salads, vegetable dishes, potatoes, and meats. Her recipe inspired my own, which I love rubbed into potatoes before roasting them. The salt is exuberantly flavorful, and is excellent on nearly all potato dishes, including scalloped potatoes, as well as on grilled and roasted poultry, salmon, and a variety of other dishes.*

2 teaspoons whole caraway seeds
1 ½ teaspoons whole coriander seeds
½ teaspoon fenugreek seeds
1 teaspoon whole black peppercorns
3 tablespoons sel gris or other unrefined solar-dried salt
Pinch of cayenne pepper

Grind the seeds and peppercorns in a suribachi or other hand grinder. Add the salt and cayenne, and mix thoroughly. Stored in a cool cupboard, this salt will last for several weeks.

Variation:

For a more fragrant, earthy salt, grind the spices with 3 crushed garlic cloves to make a paste before adding the salt. Use within 2 days.

Cajun Spice Mixture

Makes about 1 ¼ cups

If you cook a lot of New Orleans and Cajun style foods, it makes sense to make your own Cajun Spice Mixture. You can many find commercial blends, typically called Creole Seasoning, which is a bit of a misnomer, as Creole cuisine is traditionally much less spicy and more refined than Cajun food. To make your own, buy spices in bulk so that you can get just what you need without spending a lot of money and having a lot of extra jars around.

3 tablespoons sweet Spanish paprika
2 tablespoons smoked Spanish paprika
2 tablespoons kosher salt
2 tablespoons onion powder
2 tablespoons garlic powder
2 tablespoons dried basil
2 tablespoons dried oregano
2 tablespoons dried thyme
1 tablespoon celery seed
1 tablespoon mustard flour, such as Colman's Dry Mustard
1 tablespoon ground cayenne
1 tablespoon ground black pepper
1 tablespoon ground white pepper

Put all of the ingredients into a wide-mouth pint glass jar, secure the lid and shake until evenly mixed.

Store in a cool, dark cupboard for up to 6 months.

Mixed Peppercorns

Makes about 1 ¼ cups

Some cooks keep several pepper grinders at the ready at all times: one for black peppercorns, one for white, one for mixed peppercorns, and one with peppercorns mixed with allspice berries or coriander. I confess to having five myself, even though I'm as likely to crush my peppercorns by hand in a suribachi or molcajete as I am to use a grinder. A grinder is handy at the table, though, and for a final sprinkling over a finished dish. Keep in mind that if you use a large peppercorn such as Tellicherry, some will inevitably get stuck in the grinding mechanism. I find this happens frequently, even with my best mills. The problem is almost always solved by whacking the grinder against a hard surface; occasionally I empty out the grinder and start over with a new batch of peppercorns.

This mixture offers a full spectrum of flavor. Use it as you would plain black or white pepper, when you want an extra burst of peppery taste.

¼ cup (½ ounce) Sichuan peppercorns, toasted, crushed, and sifted
¼ cup (1 ounce) white peppercorns
½ cup (2 ounces) black peppercorns
¼ cup (½ ounce) dried green peppercorns
1 tablespoon (⅛ ounce) whole coriander
1 tablespoon allspice berries, coarsely cracked

Mix all of the spices together in a bowl, tossing gently to distribute them evenly. Fill a pepper grinder with the mixture and store any leftover mixture in a glass jar in a dark pantry and use to refill the grinder.

Lagniappe

"Here's a little something extra for you," a clerk in Haddiesburg, Mississippi, or Shrevsport, Louisiana, may say as he tucks a treat of some sort into your bag. It's the thirteenth doughnut, the extra tomato, the bunch of basil, an additional peach or two.

The word itself is French and so in the United States we find it in that most French part of the country, New Orleans and its neighbors.

The idea of a little something extra has expanded to mean all manner of small unrequested treasures, from something you can pop in your mouth to advice you may not have known you needed but that proves to be invaluable. Here, I offer a few inedible gems, my gifts to make life a little more fun using my favorite ingredients, salt and pepper.

Playdough, Uncooked
Playdough, Cooked
Salt Dough Decorations
Pepper-Scented Paper

Playdough, Uncooked

Makes 2 6- to 8-inch balls

Playdough is a great way to keep kids occupied and you don't need to buy those little containers of it, as it is very easy and inexpensive to make at home. This version comes from my former assistant, Lesa Tanner, who was working for me as I was writing the first edition of this book. The mother of three energetic boys, I took her word on the effectiveness of playdough, especially during rainy days and summer vacation.

5 cups all-purpose flour
1 ½ cups salt
6 tablespoons vegetable oil
3 cups boiling water, with food coloring of choice added
6 tablespoons alum, available at pharmacies

Combine the flour and salt in a large mixing bowl. Add the boiling colored water, the oil, and the alum. Mix together, turn out onto a floured board, and knead until the dough is no longer sticky. Once the dough is cooled, store in a heavy-duty plastic bag.

The dough will keep for several months if stored in a sealed bag between uses.

A Salt & Pepper Cookbook

Playdough, Cooked

Makes 1 6- to 8-inch ball

To color dough with natural ingredients rather than commercial food coloring, use water in which you have boiled yellow onion skins, red beets, golden beets, or, for blue, yellow onion skins and red or golden beets. For other native plant dyes, visit the USDA Forest Service website at fs.fed.us/wildflowers.

2 cups all-purpose flour
1 cup table salt
4 teaspoons cream of tartar
2 cups water
4 tablespoons vegetable oil
10 to 15 drops of food coloring (or more, as desired)

Combine all the ingredients in a large, heavy pot set over medium heat. Cook, stirring constantly, until the dough comes together in a ball. Turn out onto an unfloured work surface and let cool. Knead the cool dough several times, then place it in a heavy-duty plastic bag. The dough will keep for several months if stored in a sealed bag between uses.

Salt Dough Decorations

Makes about 2 dozen 2 ½-inch decorations

John Ash, a world-renowned chef, cookbook author, and culinary teacher, lives in Sonoma County and is a good friend. Although he is widely know for his culinary expertise and success, there's one charming little part of his past that is largely unknown. He and his then-wife supported themselves making Christmas ornaments out of dough. John has a Bachelor's Degree in Fine Arts, with painting as his specialty, a talent he employed to craft their whimsical holiday designs.

No Christmas tree is complete without a few salt dough decorations, though certainly the dough's decorative possibilities are not limited to the winter holidays. Halloween ornaments, Valentine hearts, and almost anything else that inspires your imagination, even zombies, can be shaped and glazed.

4 cups all-purpose flour
1 cup table salt or 1 ¼ cups kosher salt
1 ¾ cups water
1 egg white
2 tablespoons water

Put the flour and salt in the bowl of a mixer fitted with the dough hook and mix briefly. Continue to operate while slowly adding the water. Knead for 5 minutes by machine or 15 minutes by hand, turn out onto a floured work surface, and roll out until the dough is ¼ inch thick. Use cookie cutters to create shapes or make your own patterns using cardboard and then cut them out using a small, thin knife.

For ornaments, use an ice pick to poke a hole in the top of each ornament before baking it.

Combine the egg white with the water; using a pastry brush, brush the surface of each shape with water. For wreaths, set the shapes in an overlapping circle on a greased baking sheet and press down lightly to seal. Bake the decorations in a 300-degree oven until hard and lightly browned, about 1 hour. Let cool and either decorate with paint or leave natural.

Coat with a clear sealant such as craft glaze.

Pepper-Scented Paper

Makes 1 8- by 11-inch sheet of paper

Anandan Abdullah, general manager of the Pepper Marketing Board of Sarawak, Malaysia, when I visited in 1998, told me one afternoon that you can make paper using the plant material left after the peppercorns have been stripped off.

"If you add some crushed peppercorns," he said, "it will remain fragrant for a long time."

I was immediately inspired to try it at home and the results of my first effort still sits on my desk, sturdy, pretty, and still mildly fragrant.

Making homemade paper is a wonderful activity for children, especially during holidays and vacations. It is, of course, important that each child end up with his or her own finished product; this recipe can be doubled or tripled as necessary, though you should process each sheet individually. You will need one or more small portable window screens (9 by 12 inches is the ideal size), a large roasting pan and a baking sheet for each screen, new sponges, and plenty of cheesecloth on hand before you begin. (If you don't have a roasting pan of the proper size, you can let the paper drain over the sink.) The paper, beautiful, fragrant, and heavily textured, is decorative, being too thick to be used for writing.

1 tablespoon grated citrus zest (orange, grapefruit, lemon, or lime), optional
3 cups, packed, small pieces of paper (see Note on next page)
1 tablespoon dried flower petals, such as roses, herb flowers, lavender, marigolds, or daisies
2 tablespoons freshly cracked black peppercorns

Let the zest dry in the air for at least an hour.

Put the paper into a large bowl and pour 4 cups of hot water over it. Let it soak until very soft, at least 1 hour or longer, depending on the texture of the paper.

When the paper is very, very soft, tip the entire mixture into the work bowl of food processor or blender and add the zest, flowers, and peppercorns. Run the processor steadily at high speed for 1 minute; stop, stir the mixture, and operate for another minute, until a smooth liquid is formed.

Set a window screen over a roasting pan slightly larger than the screen, pour the paper mixture over the screen, and use your hand to distribute the tiny bits of paper and herbs evenly. Let the water drain into the roasting pan for 10 minutes, pressing it down occasionally to squeeze out more water.

Cut a triple-thick piece of cheesecloth to fit over the screen, set it on top of the wet mixture, and then press down to continue to remove excess water.

Set a sheet pan bottom-side down over the cheesecloth and turn the whole affair over, so that the cheesecloth and the wet paper are on top of the bottom of the sheet pan. Carefully lift off the screen so that the cheesecloth and paper remain on the sheet pan.

Use a clean, new sponge to press the paper down and absorb excess water. Make another triple-thick piece of cheesecloth and press it on top of the paper, pushing down evenly and firmly. Let sit in a cool dry place until completely dry, It will take from 1 to 5 days, depending on temperature and humidity. When it is thoroughly dry, peel off both layers of cheesecloth and use as you like.

NOTE

Construction paper, artist's drawing paper, wrapping paper, thin paper bags (not standard brown shopping bags) are all suitable. Tear or cut the paper into small pieces, or the bits may be caught in the blade of your processor or blender. For the best results, coordinate colors too, choosing a selection of papers of similar colors, shades of yellow and gold; shades of blue; shades of red, etc.

PART IV
Appendices

Tasting Notes & Recommendations

Salts

So much has changed in the world of salt since I wrote the first edition of this book that it is no longer possible or practical to identify all the salts you will find in the marketplace, especially the specialty marketplace. Many salts are repackaged under company names and brand names. In addition, many salts are now manufactured, adding charcoal to create black salt that is then identified as "lava salt" and adding red clay to create a deep orange salt that looks like traditional Hawaiian alaea salt but, in many cases, has never been near the islands. Hence, I have changed this section to identify the most readily available, the most reliable, and the ones I recommend.

Flake Salt: All flake salts are produced similarly, as it takes specific conditions—a heated saturated brine—for salt to form a flake. All flake salts dissolve quickly; for general cooking, price should be a guide.

Name	Source, Price	Taste	Best Uses
Cyprus Mediterranean	Cyprus; seawater; expensive	Delicate; bright; smooth	Finishing
Hana No Shio	Japan; seawater; expensive	Delicate; bright; smooth, with snap	Finishing
Kosher (Diamond Crystal brand)	U.S.; mined; inexpensive	Smooth; delicate; dry	All-purpose cooking, finishing
Maldon	England; seawater; moderate	Smooth; delicate; dry	Finishing
Murray River	Australia; underground salt water deposits; moderate	Light; delicate; smooth; dry	Finishing

Solar Dried Salt, Unrefined

The best solar dried sea salts are those that go through no additional processing.

Name	Source, Price	Taste	Best Uses
Celtic Sea Salt (brand name)	Atlantic coast, France; moderate	Complex; earthy; minerally; moist	Finishing, general cooking
Fleur de Sel	Guerande, France; seawater; expensive	Delicate; bright; moist	Finishing
Sel Gris	Atlantic coast, France; seawater; moderate	Complex; earthy; minerally; moist	Finishing, general cooking
Sel de Guerande	Brittany, France; seawater; moderate to expensive	Complex; earthy; minerally; moist	Finishing, general cooking

Solar Dried, Refined

Many sea salts—i.e., salt from evaporated seawater—go through additional processing that removes trace minerals and typically creates a small hard cube. These salts may be labeled as sea salt, table salt, and, if iodine is added, iodized salt; they often come from the same pile of salt, regardless of their label descriptions. These salts are sometimes called granulated.

Name	Source, Price	Taste	Best Uses
Baleine	Mediterranean coast of France; moderate	Sharp; hard	N/A
Kosher Salt	U.S.	**Sharp; hard; slow to dissolve**	**General cooking where flake salt is not available**

(Continued)

Name	Source, Price	Taste	Best Uses
Sea Salt (sold under many brand names)	U.S.; inexpensive	Sharp; hard	N/A
Table Salt	World wide; inexpensive	Sharp; hard	Non-food uses, such as cleaning
Table Salt, Iodized	World wide, inexpensive	Sharp; hard	General cooking where iodine is not readily available in foods

Mined, Unrefined

All salts may correctly be called sea salt, as all come from seawater, some of it now as ancient inland salt caverns, the most famous of which is currently Himalayan pink salt.

Name	Source, Price	Taste	Best Uses
Himalayan	Pakistan; moderate to expensive	Bright; metallic; hard	Finishing; in salt grinders/graters; salt plates, blocks & bowls; rooms made of salt
Indian Black	India; cheap	Strong taste of sulphur, with bass note of salt	Traditional Indian dishes
Utah Red	Utah; inexpensive to moderate	Iron deposits give this salt a taste of blood	Novelty

Manufactured

This category of salts includes those that have ingredients other than sodium and chloride in the flakes or crystals. The most common is Hawaiian alaea, which is manufactured to resemble traditional gathered alaea salt, which cannot be sold because it may contain dirt, seaweed, and other items. Black salts are typically called "lava salt" because of their color.

Black Salt	Various; moderate to expesnive	Moderately salty, with a taste of carbon; both flake and rock	Novelty
Hawaiian Alaea Salt	Various; inexpensive to moderate	Moderately salty; the best have a silky feel on the palate; some taste of iron or blood	Traditional Hawaiian cooking and ceremonies
Smoked Salt	Various; moderate	Moderately salty; aggressively smoky; the best are flake salts; some are hard	Finishing, especially with eggs and cheese

Peppercorns

Although peppercorns are produced in more countries than are listed here, most pepper sold in the United States comes from the countries mentioned. The source of pepper is not always included on a retail label; when it is, it is typically because there is a perception of superior quality associated with the location. Malaysian peppercorns are currently very rare in the United States but I have kept them in this list, as I believe they are the best or among the best peppercorns in the world.

Name	Source	Description
Creamy White Pepper (brand name)	Sarawak, Malaysia	White, consistently sized and colored; medium-large berries; superior quality
Brazilian	Brazil	Black, white, or green; medium berries; average quality
Green, air-dried	Various	Medium berries; average to superior quality
Green, freeze-dried	Various	Small, light (weight) berries; very fragile; fair quality
Green, in brine	Various	Soft, medium to large berries; average to superior quality
Green, in vinegar	Various	Soft, medium to large berries; average to superior quality
Lampong	Indonesia	Black; medium berries; average quality
Malabar	India	Black; medium to large berries; average to superior quality
Muntock	Indonesia	White; medium berries; average quality
Naturally Clean Black Pepper* (brand name)	Sarawak, Malaysia	Black; medium berries; superior quality; floral, almost sweet aromas; rare and special
Ponape (brand name)	Micronesia	Black; medium berries; average to superior quality; very limited production
Sarawak	Sarawak, Malaysia	Black or white, generic; average to superior quality
Tellicherry	India	Black, frequently with reddish cast; very large berries; superior quality
Vietnamese	Vietnam	Black with reddish tones; medium berries; average to good quality

Resources
for Salts, Peppercorns, Accessories, Information, & Really Good Bacon

American Spice Trade Association (ASTA)
1101 17th St., NW Suite 700
Washington, DC 20036
Website: astaspice.org
Tel: 202-331-2460
Consumer and industry information about spices, including black, white, and green peppercorns

Aquasel
10 rue des Marouettes
85330 Noirmoutier
France
Tel: 33.2.51.39.08.30
Website:aquasel.fr
Cooperative of regional salts

British Salt Limited
Cledford Lane
Middlewich
Cheshire CW 10 OJP
England
Tel: 44 0 1606 839 250
Website: british-salt.co.uk
Information on the history and production of mined salt in England

The Cat Museum
Kuching North City Hall
Bukit Siol, Jalan Semariang, Petra Jaya
93050 Kuching, Sarawak
Malaysia
Tel: (60) 082-446-688
Website: sarawaktourism.com
One of the world's only cat museums, in one of the world's major pepper producing regions

Corti Brothers
P.O. Box 191358
Sacramento, CA 95819
Tel: (916) 736-3800
Website: cortibrothers.com

Japanese sea salt, Trapani salt, other
specialty salts as well as olive oils,
vinegars, etc.; newsletter

The Grain & Salt Society
273 Fairway Drive
Asheville, NC 28805
Tel: (888) 353 0030
Website: celticseasalt.com
Celtic Sea Salt brand French sea salts,
other salts, organic peppercorns

International Pepper Community
4th Floor, Lina Building
JL. H.R. Rasuna Said Kav. B7
Kuningan, Jakarta
Indonesia
Tel: +62-21 522-4902
Website: ipcnet.org

Kalustyan's
123 Lexington Avenue
New York, NY 10016
Tel: (800) 352-3451
Website: kalustyans.com
Quality spices and more

La Cuisine—The Cook's
Resource
323 Camereon Street
Alexandria, VA 22314-3219

Tel: (800) 521-1176
Website: lacuisineus.com
French, English, and Spanish sea salts;
peppercorns; grinders, including the
electronic grinder, A Touch of Pep-
per; newsletter, mail order

La Coopérative des Sauniers
de l'Île de Ré
7 Route de la Prée
Ars-en-Ré, France
Tel: + 33 05 46 29 40 27
Website: sauniers–iledere.com
Salt on the Île de Ré

La Maison des Paludiers
et Le Groupement des
Producteurs de Sel de La
Presqu'île Guérandaise
18, Rue des Prés Garnier
44350 Guérande
France
Website: seldeguerande.fr
Information on purchasing the region's
famous salts, visiting the salt
marshes and more

Local Spicery
80-F Main Street
Tiburon, CA 94920
Tel: (415) 435-1100

Website: http://www.fresh
localspices.com/
Purveyors of freshly milled spices, whole
spices, including peppercorns, signa-
ture blends, salts, and more, including
a spice-of-the-month club. A retail
location opens in late spring, 2015.

Malaysian Pepper Board

P.O. Box 1653
93916 Kuching, Sarawak
Malaysia
Tel: 082-331811
Email: pmb@pepper.po.my
Website: mpb.gov.my
Processors of Naturally Clean Black
Pepper, Creamy White Pepper, and
other peppers for the world market

Maldon Crystal Salt Company Limited

Wycke Hill Business Park
Maldon, Essex, England, CM9 6UZ
Tel: +44 (0)1621 853 315
Website: www.maldonsalt.com
Consumer information on the history and
production of sea salt in England

The Meadow

523 Hudson St.
New York City, 10014

Tel: (212) 645-4633
Website: atthemeadow.com
Purveyors of over 100 specialty salts,
salt blocks and bowls and more, with
three locations, two in Portland, Ore-
gon, and this one. Home of America's
first selmelier, Mark Bitterman.

Onondaga County Salt Museum

Onondaga Lake Park
106 Lake Drive
Liverpool, NY 13088
Tel: (315) 453-6715
Website: onondagacountyparks.com
History of the salt industry in Syra-
cuse, with artifacts, equipment,
photographs, replica of production
facilities, a gift shop, picnic grounds,
and free admission.

Pepper-Passion

1111 East Madison St., #124
Seattle, WA 98122
Tel: (877) 658-3373
Website: pepper-passion.com
Purveyors of peppercorns from a wide
selection of producers, including
Penja Black, a lovely pepper from
Cameroon, Africa, and a small line

of pepper grinders, including hand-carved and signed specialty mills of African olive wood

R. M. Felts Packing Company

Box 199
Ivor, VA 23866
Tel: (757) 859-6131
Producers of authentic smoked bacon, hog jowls, and country ham

Royal Wieliczka Salt Mine

Kopalma. Soli Wieliczka
ul. Danilowicza 10
32-020 Wieliczka, Poland
Tel: (4812) 278-73-02
Website: wieliczka-saltmine.com
Wieliczka salt mine; the Muzeum Zup Krakowskich Wieliczka (a salt museum) is located at the mines with a separate entrance (and additional entry fee)

Salt Institute

700 North Fairfax Street
Fairfax Plaza, Suite 600
Alexandria, VA 22314-2040
Tel: (703) 549-4648
Website: www.saltinstitute.org
Consumer information about salt and related health issues

Salt Traders, LLC

18D Mitchell Rd.
Ipswich, MA 01938
Tel: (800) 641-7258
Website: salttraders.com
Purveyors of specialty salts, salt blocks, plates and bowls, bulk salts, salt licks, peppercorns, spice blends, spice grinders, and more

SaltWorks

16240 Wood-Red Rd. NE
Woodinville, WA 98072
Tel: (800) 353-7258
Website: saltworks.us
Purveyors of specialty salts, bulk salts, bath salts, flavored salts, and "Sonoma Sea Salt"

Strataca Kansas Underground Salt Museum

3650 East Avenue G.
Hutchinson, KS 67501
Tel: (866) 755-3450
Website: underkansas.org
Information on the salt industry that was founded in Hutchinson in 1875

Tabasco Country Store
McIlhenny Company
Highway 329
Avery Island, LA 70513
Tel: (800) 634-9599
Website: countrystore.tabasco.com
The famous Louisiana hot sauce
company shares Avery Island with
one of the country's oldest salt
mining companies; catalogue

Zingerman's
422 Detroit Street
Ann Arbor, Ml 48104
Tel: (888) 636-8162
Mail-order source for specialty salts,
cheese, vinegars, etc.; catalogue

Glossary

Alaea salt • *See* Hawaiian sea salt.

Alberger salt • A method of producing hollow, pyramid-shaped crystals of pure salt developed by J. L. Alberger and patented in 1889. Alberger salt has replaced grainer salt; the largest crystal is sold as Diamond Crystal Kosher Salt; four other sizes are sold under various names, and primarily to the food service industry.

Bittern • The bitter liquid that remains after salt has been made from seawater or from dissolved rock salt. In some countries, but not in the United States, where the practice is prohibited, bittern is drained into the sea. Here it is sold for various industrial uses, including dust control on state and national park roads.

Black peppercorns • The berries of the *Piper nigrum* vine, picked while green and dried either in the sun or in forced-air dryers.

Black pepper oil • Volatile oils responsible for the aroma of black pepper; used in small quantities in therapeutic massage (to relieve muscle stiffness) and in aromatherapy as a stimulant and an aphrodisiac.

Black salt *(Kala namak)* • Sometimes called rock salt (but not what is called rock salt in the United States), black salt ranges in color from pale violet to purple black; it has a strong sulfuric aroma and is used almost exclusively in regional Indian cooking.

Black soy sauce • *See* Soy sauce.

Brine • A solution of salt dissolved in water; a saturated brine is 26.4 percent salt by weight; at higher percentages, salt will not remain in solution and begins to precipitate out, forming fragile crystals on top of the liquid that quickly collapse from their own weight and sink.

Celtic Grey Sea Salt • A trademarked name for one brand of sel gris From the Atlantic coast of France (see sel gris).

Condiment salt • A contemporary name for salts that should be sprinkled onto food immediately before serving rather than used during cooking.

Creamy White Pepper • A high grade of white pepper, with all the dark-colored and lightweight berries sorted out; produced by the Pepper Marketing Board of Sarawak, Malaysia.

Cubeb • A small berry *(Piper cubeha),* native to Southeast Asia, related to *P. nigrum* but more closely resembling all-spice or nutmeg than pepper; gathered in the wild, cultivated in limited areas, and little used outside Asia.

Danish salt • *See* Smoked salt.

Dark soy sauce • *See* Soy sauce.

Dendridic salt • Small hollow salt crystals (the size of table salt), made by introducing yellow prussiate of soda into a saturated brine; drawbacks include the small size (larger crystals break apart) and limited uses in the food industry because it forms a blue compound in the presence of iron.

Fagara • *See* Sichuan peppercorns.

Finishing salt • *See* Condiment salt.

Fish sauce • A clear, brown liquid made by fermenting anchovies with sea salt and water. The most important seasoning in Thai cuisine *(nam pla)* and Vietnamese cuisine *(nuoc mam).*

Flavor salt • Monosodium glutamate, used as a flavor enhancer in many Asian cuisines, and in Asian countries often labeled as "flavor salt" or "salt with flavor."

Fleur de sel **(flower of the sea)** • The queen of the sea salts, produced along the French Atlantic coast; the top crust of salt in the salt pans in Brittany. Once discarded as unprofitable, it is now the world's highest-priced salt and the favorite of many chefs. It is delicately flavored, but it's main contribution to a dish is its texture. Too expensive for an all-purpose salt; use to finish a dish.

Flower pepper • *See* Sichuan peppercorns.

Fresh green peppercorns • The freshly picked clusters of green peppercorns before they are dried or further processed; used in a few classic dishes, such as a Thai green curry; almost impossible to find except locally where pepper is grown.

Gomashio • A Japanese condiment of toasted sesame seeds crushed with salt and, sometimes, dried seaweed.

Grainer salt • Evaporated salt produced using heat-stimulated evaporation (originally, from direct heat under the pans; later, from steam pipes directly in the brine). Although the process created a desirable crystal form (a hollow pyramid), grainer salt is no longer produced commercially because of the cost of the heating fuel.

Grains of Paradise • Seeds of a perennial reed, cultivated in west Africa; a relative of cardamom, it is milder, with a pungent and peppery flavor; sometimes called Guinea pepper.

Granulated salt • *See* Table salt.

Green peppercorns • Pickled, freeze-dried, or air-dried peppercorns picked several weeks before white and black peppercorn berries are harvested. Fresh-tasting, mildly sour, and peppery in flavor; widely available.

Hawaiian alaea salt • A pale orange salt made with Hawaiian red clay; the commercial equivalent of traditional Hawaiian red salt, which includes significant quantities of soil and other ingredients and is not approved by the Food and Drug Administration for human consumption.

Indian salt • *See* Black salt.

Industrial salt • A general term that refers to all non-food-grade salt and which makes up over 90 percent of the salt used in the United States each year.

Iodized salt • Table salt to which potassium iodide has been added as a preventive against goiter (a thyroid condition).

Japanese pepper • *Sansho,* the ground pod of the Japanese prickly ash tree *(Zanthoxylum piperitum),* is occasionally referred to as "Japanese pepper" although it is not hot and not related to *P. nigrum*. Rather, it is tangy and the ground spice is sprinkled on food at the table as a contrast to fatty flavors, especially those of grilled poultry and seafood.

Japanese sea salt • Generic name for sea salt from Japan.

Kala namak • *See* Black salt.

Korean sea salt • Generic name for sea salt from Korea.

Kosher salt • Virtually without exception, what is known as kosher salt is a coarse crystal that is efficient at withdrawing liquid from meat, the most crucial part of koshering.

Ksosian sea salt • Generic name for salt from the Ksosian area of South Africa.

Lampong pepper • The primary black pepper from Indonesia, now shipped out of the port of Pendjang; Lampong is on the island of Sumatra; Krakatoa, formidable and steamy, looms across the bay.

Light soy sauce • *See* Soy sauce.

Lima sea salt • Brand name for a type of sea salt produced along the Atlantic coast of France.

Liquamen • A seasoning, used in Roman times, in place of salt; similar to a crude fish sauce, its simplest versions consisted of fish dried in the sun and combined with salt, water, and various seasonings; also known as *garum.*

Long pepper • *Piper Longum,* from India, has a peppery taste, with spicy cinnamon notes; once more common than *P. nigrum,* it is rarely found today outside India and Indian markets.

Malabar pepper • A medium-sized, intensely flavored variety of *P. nigrum* from India; said by many to have the best taste among black peppercorns.

Maldon Salt Flakes • Coarse, uneven flake salt from Essex, England, and widely available in the United States. Because of its thin, flaky texture that melts quickly on the tongue, it is an excellent finishing salt. Now offered smoked, as well.

Miso • Fermented bean paste, essential in Japanese cooking. There are several types of miso, nearly all made with a mixture of crushed boiled soybeans and grain (wheat, barley, or rice) that is inoculated and allowed to ferment. Most are salty; a few are sweet; some are sweet-sour.

Muntock pepper • White peppercorns from Indonesia.

Naturally Clean Black Pepper • Premium-grade black pepper of consistent quality produced by the Pepper Marketing Board of Malaysia; unlike most black pepper on the world market, NCBP is washed twice immediately after picking to remove bacteria and then processed in air-dryers rather than dried in the sun (where it is exposed to both salmonella and *E. coli).* It is sterilized using steam and has the lowest bacteria count of all black pepper. Considered by some, including myself, to be the best black pepper in the world.

Oshima Island Red Label Salt • Brand name of the top grade of salt produced on Ö-shima Island, Japan.

Paulidier • A salt farmer, the term used along the Atlantic coast of France, as is *saunier.*

Pepperberry • A spicy berry, about twice the size of a peppercorn from a shrub *(Schinus molle)* native to Tasmania; common in Australia.

Pepperleaf • Native to Australia, the dried leaf has a mildly spicy flavor reminiscent of black pepper; ground, it is used as a seasoning in Australia.

Pickling salt • Pure food-grade salt, without additives (such as anticaking agents) that will cloud liquid.

Pink peppercorns • Not a true peppercorn, these pink berries from a bushy tree *(Schinus terebinthifolius)* that grows in Florida, Brazil, and the island of Réunion, are colorful, sweet, have just a hint of mildest pepper flavor, no heat, and a lengthy bitter finish. More of a novelty item than a useful ingredient.

Piperine • The active ingredient in pepper responsible for its heat and pungency; does not dissipate when heated or over time. Can be extracted as *piperine oleoresin,* which is used in both the food and beverage industries.

Piper **spp.**

> *P. grande* • A variety of wild pepper, with red berries.

> *P. longum* • *See* Long pepper.

> *P. nigrum* • *See* Black peppercorns; Green peppercorns; White peppercorns.

> *P. vestitum* • A variety of wild pepper, with golden brown and black berries.

Popcorn salt • Very fine salt that clings readily to popped corn.

Portuguese salt • Generic name for sea salt from Portugal.

Red salt • *See* Hawaiian alae salt.

Rock salt • Chunk salt that does not meet the Food and Drug Administration's requirements for food-grade salt; generally used in ice cream freezers and for melting winter ice.

Salt beef • Salted and dried beef, popular in Caribbean and Cuban cooking; in England, salt beef is similar to corned beef.

Salt fish • Salted and dried fish; cod is the most popular and most common, but various other fish are dried in this way.

Salt with flavor • *See* Flavor salt.

Salt pork • Pork belly cured in brine; not smoked.

Salt-Sense • Brand name for medium-sized Alberger salt crystals, slightly smaller than kosher salt. Less dense than granulated salt and so, by volume, it contains less sodium chloride; advertised as an alternative to table salt.

Salt substitute • Usually potassium chloride in granular form marketed to people who want to lower their sodium intake; slightly bitter and acidic in taste.

Salumi • Salt-cured meats; Italian.

Sansho • *See* Japanese pepper.

Sarawak pepper • Generic name for peppercorns processed by the Pepper Market Board from peppercorns grown in the Malaysian state of Sarawak, on the island of Borneo.

Sea salt • Any salt made from evaporated seawater; much table salt is also sea salt and many sea salts are, as table salt is, completely refined.

Sel gris • Generic name for gray sea salt from the northwest coast of France.

Shrimp paste • A pungent fermented paste made from salt and shrimp, essential in Southeast Asian cuisines *(kapi* in Thai, *blachan* in Malay, and *terasi* in Vietnamese). There are two types: fresh, which needs no preparation, and dried, which should be wrapped in foil and roasted in a dry skillet before being used.

Sichuan pepper • An aromatic reddish-brown berry from the Chinese prickly ash tree *(Zanthoxylum simulans),* sometimes called flower pepper and more often, fagara. Used toasted in Chinese cooking (it's essential in Chinese five-spice powder) and combined with salt, as a condiment.

Sicilian salt • Generic name for sea salt from Sicily.

Smoked salt • Sea salt that is smoked over wood; made in Denmark.

Soy sauce • A salty fermented condiment with a thousand-year history that has become one of the most essential seasonings throughout Asia. Made from roasted soybeans, wheat, salt, and a culture, the best is fermented for a year or more. The least expensive, and least desirable, are synthesized from vegetable protein, hydrochloric acid, caramel coloring, and corn syrup.

Dark soy sauce • In Japanese cooking, dark soy sauce is the standard soy sauce, nearly black in color, with substantial body and less saltiness than light soy sauce. Most Chinese soy sauces, which are less common than the Japanese sauces are in this country, are thick and dark and often called black soy sauce.

Light soy sauce • Traditionally, light soy sauce is one of two types of soy sauce generally used in Japan. It is thinner and lighter in color than is dark soy sauce, but it is also saltier.

Suribachi • A Japanese mortar and pestle, consisting of a deep bowl of ridged ceramic and a long wooden pestle. It's inexpensive and highly effective for crushing salt and peppercorns, as well as for grinding garlic, chiles, and herbs into pastes quickly and efficiently.

Table salt • Small, uniform cubes, also called granulated salt, usually iodized.

Tamari • Similar to soy sauce, but made without wheat.

Tellieherry pepper • The name once referred to pepper shipped through the port of Tellieherry in southern India, but now indicates a grade of Indian black pepper: the larger, more mature peppercorns with a complex, round flavor.

Virgin pickle • A 100-percent saturated brine of evaporated seawater or dissolved rock salt; the last stage of evaporation before natural crystallization begins.

Volatile oils • The oils responsible for the aroma of pepper; they dissipate quickly when heat is applied and when pepper is crushed or ground, leaving behind piperine, which accounts for the lasting heat of pepper.

White peppercorns • Peppercorns picked shortly (a week or less) after the berries for black pepper are harvested, and soaked in water for ten to fourteen days to remove the skins before being dried; less aromatic than black pepper, but often hotter, due to a slightly higher piperine content.

Bibliography

Aikman-Smith, Valerie. *Salt: Cooking with the World's Favorite Seasoning.* New York: Ryland, Peters & Small, 2009.

Alderman, Michael H., and Bernard Lamport. "Moderate Sodium Restriction: Do the Benefits Justify the Hazards?" *American Journal of Hypertension* 3 (1990): 499–504.

Alford, Jeffrey, and Naomi Duguid. *Flatbreads and Flavors.* New York: Morrow, 1995.

Aragon, Jane Chelsea. *Salt Hands.* New York: Puffin Unicorn, 1989.

Barron, Rosemary. *Flavors of Greece.* New York: Morrow, 1991.

Behr, Edward. *The Artful Eater.* New York: The Atlantic Monthly Press, 1992.

Bharadwaj, Monisha. *The Indian Spice Kitchen.* New York: Dutton, 1997.

Bitterman, Mark. Salted: *A Manifesto on the World's Most Essential Mineral, with Recipes.* Berkeley: Ten Speed Press, 2010.

———. *Salt Block Cooking.* Kansas City: Andrews McMeel, 2013.

Bosker, Gideon. *Great Shakes: Salts and Peppers for All Tastes.* New York: Abbeville Press, 1986.

Bremness, Lesley, and Jill Norman. *The Complete Book of Herbs & Spices.* New York: Viking Studio Books, 1995.

Carey, Benedict. "Shake-Up in the Saltshaker." In *Health* (May/June 1990): 24.

Carey, Larry, and Sylvia Tompkins. *1002 Salt Shakers: Nodders, Fitz & Floyd, Parkcraft with Prices.* Atglen: Schiffer Publishing, Limited, 1995.

Chan, Dunstan, and Woon Yoke Heng, eds. *Sarawak Pepper Flavors the World.* Malaysia: Pepper Marketing Board Malaysia, 1994.

Collins, Larry, and Dominique Lapierre. *Freedom at Midnight.* New York: Simon and Schuster, 1975.

Colwin, Laurie. *More Home Cooking.* New York: HarperCollins, 1993.

David, Elizabeth. *Spices, Salt and Aromatics in the English Kitchen*. Middlesex, England: Penguin, 1970.

de Groot, Roy Andries. *Feasts for All Seasons*. New York: Knopf, 1966.

Devi, Yamuna. *The Vegetarian Table: India*. San Francisco: Chronicle Books, 1997.

"Does Salt Raise Your Blood Pressure?" *Consumer Reports on Health* (April 1994): 42–43.

Eskew, Garnett Laidlaw. *Salt: The Fifth Element*. Chicago: Ferguson, 1948.

Evans, John, and Lynette Evans, "A Pinch of Salt." *San Francisco Examiner* (February 25, 1998).

Field, Carol. *Italy in Small Bites*. New York: Morrow, 1993.

Fussell, Betty, and Michele Anna Jordan. "Savoring Salt." *Bon Appetit* (March 1998): 58.

Gelle, Gerry G. *Filipino Cuisine*. Santa Fe: Red Crane Books, 1997.

Glynn, David and Fritz Gubler. *The Salt Book: Your Guide to Salting Wisely and Wall, with Recipes*. Crows Nest, Austrailia: Arbon Publishing, 2010.

Gugino, Sam. "Worth Its Salt," *The Wine Spectator* (15 September 1998): 25–26.

Hachten, Harva. *Kitchen Safari*. New York: Atheneum, 1970.

Hamlin, Suzanne. "Salt Is Regaining Favor and Savor." *New York Times* (5 June 1996): Cl.

Hanneman, Richard L. "The Politics of Sodium Restriction in the United States." In *Seventh Symposium on Sait,* i vols., ii–iv 1993, 2, 231–39. Amsterdam: Elsevier Science Publishers.

Harris, Dunstan A. *Island Cooking: Recipes from the Caribbean*. Freedom, Calif.: The Crossing Press, 1988.

Hazan, Marcella. *Marcella Cucina*. New York: HarperCollins, 1997.

Hendra, Tony. "Salty Talk." *Food & Wine* (April 1998): 55–56.

Hesser, Amanda. "From the French Marshes, a Salty Treasure." *New York Times* (6 May 1998): F1.

Hutton. Wendy, ed. *The Food of Malaysia*. Singapore: Periplus Editions, 1995.

Ibrahim, M.Y., C. F. J. Bong, and I. B. lpor. *The Pepper Industry: Problems and Prospects*. Bintulu: Universiti Pertanian Malaysia, 1993.

Jenkins, Steve, *Cheese Primer*. New York: Workman, 1996.

Jones, Evan. *The World of Cheese*. New York: Knopf, 1976.

Kamman, Madeleine. *The New Making of a Cook*. New York, Morrow, 1997.

Kaufmann, Dale W. *Sodium Chloride : The Production and properties of Salt and Brine*. New York: Hafner, 1968.

Koch, Ulrike, director. *The Saltmen of Tibet*. Digital video, 110 min. Catpics Coproductions, Zurich, Switzerland, 1998.

Kurlansky, Mark. *Salt: A World History*. New York: Penguin Books, 2002.

_____. *The Story of Salt*. New York, Penguin Books. 2006

Kyle, Calvin. *Gandhi, Soldier of Nonviolence*. Cabin John, Md.: Seven Locks Press, 1983.

Lang, Jennifer Harvey. *Tastings*. New York: Crown, 1986.

Lewallen, John, and Eleanor Lewallen. *Sea Vegetable Gourmet Cookbook and Wildcrafter's Guide*. Mendocino, Calif.: Mendocino Sea Vegetable Company, 1996.

Loha-unchit, Kasma. *It Rains Fishes*. San Francisco: Pomegranate Art Books, 1994.

McGee, Harold. *On Food and Cooking*. New York: Charles Scribner's Sons, 1984.

_____. *The Curious Cook*. San Francisco: North Point Press, 1990.

Madison, Deborah. *Vegetarian Cooking for Everyone*. New York: Broadway Books, 1997.

"The Many Reasons to Cut Back on Salt." *University of California at Berkeley Wellness Letter 2,* no. 10 (July 1995): 1–2.

Moore, Thomas. J. "Overkill." *The Washingtonian* (August 1990): 64.

Multhauf, Robert P. *Neptune's Gift: A History of Common Salt*. Baltimore: Johns Hopkins University Press, 1978.

Niman, Skip. "Salt Is Not Just Salt—Considerable Differences Exist." *Cereal Foods World* 42, no. 10 (October 1997): 808–11.

O'Neill, Molly. "All It's Cracked Up to Be." *New York Times Magazine* (12 January 1997): 41–42.

———. "Doctor Pepper." *New York Times Magazine* (19 January 1997): 53–54.

Paniz, Neela. *The Bombay Cafe*. Berkeley: Ten Speed Press, 1998.

Pepper Marketing Board of Malaysia. *Report of National Pepper Investment Seminar*. Kuching, Sarawak: Pepper Marketing Board, 1995.

Robinson, Marilynne. *Housekeeping*. New York; Farrar, Straus & Giroux, 1981.

Root, Waverley. *Food*. New York: Simon and Schuster, 1980.

Rosenblum, Mort. *Olives*. New York: North Point Press, 1996.

Severance John B. *Gandhi, Great Soul* New York: Clarion Books, 1997.

"The Shake Out on Sodium: An Interview with David A. McCarron, M.D." *Total Health* (December 1994): 46.

Sigal, Jane, *Backroad Bistros, Farmhouse Fare*. London: Pavilion Books, 1994.

Silverton, Peter. "Sneezy Spice." *The Observer Review* (London) (23 November 1997): 7.

Sim, E. S. "Pepper Industry in Sarawak." Sarawak, Malaysia: Department of Agriculture, 1990.

Steingarten, Jeffrey, *The Man Who Ate Everything*. New York: Knopf, 1997.

Stewart, Katie. *The Joy of Eating*. Owing Mills, Md,: Stemmer House, 1977.

Tisdale, Sallie. *Lot's Wife: Salt and the Human Condition*. New York: Henry Holt, 1988.

Trager, James. *The Food Chronology: A Food Lover's Compendium of Events and Anecdotes, from Prehistory to the Present*. New York: Henry Holt, 1995.

Tsuji, Shizuo. *Japanese Cooking: A Simple Art*. Tokyo: Kodansha International, 1980.

Turner, Peter, Chris Taylor, and Hugh Finlay. *Malaysia, Singapore & Brunei*. 6th ed. Hawthorne, Australia: Lonely Planet Publications, 1996.

Visser, Margaret. *Much Depends on Dinner.* New York: Grove Press, 1986.
_____. *The Rituals of Dinner.* New York: Grove Weidenfeld, 1991.

Waldron, Maggie. *Cold Spaghetti at Midnight.* New York: Morrow, 1992.

Watson, Francis. *A Concise History of India.* London; Thames and Hudson, 1974.

Welsh, Willard. *Hutchinson: A Prairie City in Kansas.* Salem: Higginson Book Company, reprint, Oct. 1995.

Whyte, Karen Cross. *The Complete Yogurt Cookbook.* San Francisco: Troubador Press, 1970.

Willan, Anne. *La France Gastronomique.* New York: Arcade Publishing, 1991.

Wolfert, Paula. *The Cooking of the Eastern Mediterranean.* New York: HarperCollins, 1994.

Wolpert, Stanley. *A New History of India.* 4th ed. New York: Oxford University Press, 1993.

Young, Gordon. "Salt: The Essence of Life." *National Geographic* 152, no.3 (September 1977): 381–401.

Zane, Eva. *Greek Cooking for the Gods.* San Francisco: 101 Productions, 1970.

Index

Index   **415**

Index   **417**

Index   **419**

641.6384 JOR
Northern Lights Library System
30800011341325
Jordan, Michele Anna, author The
good cook's book of salt and
pepper : achieving seasoned delight,
with more than 150 recipes